Modeling and Simulation in Scilab/Scicos
with ScicosLab 4.4

Stephen L. Campbell, Jean-Philippe Chancelier
and Ramine Nikoukhah

Modeling and Simulation in Scilab/Scicos with ScicosLab 4.4

Second Edition

 Springer

Stephen L. Campbell
Department of Mathematics
North Carolina State University
2108 SAS Hall
P.O. Box 8205
Raleigh NC 27695
USA
slc@math.ncsu.edu

Jean-Philippe Chancelier
Ecole Nationale des Ponts et Chaussées
Centre d'Enseignement et de Recherche en
Mathématiques et Calcul Scientifique
6-8 avenue Blaise Pascal
77455 Marne-la-Vallee
Cité Descartes, Champs sur Marne
France
chancelier@cermics.enpc.fr

Ramine Nikoukhah
INRIA
Rocquencourt
78153 Le Chesnay Cedex
France
ramine.nikoukhah@inria.fr

ISBN 978-1-4939-3868-1   ISBN 978-1-4419-5527-2 (eBook)
DOI 10.1007/978-1-4419-5527-2
Springer New York Dordrecht Heidelberg London

MSC 2010: 68U20, 68U07, 65-01, 37M05, 00A72, 65Y15, 34H05

Springer is part of Springer Science+Business Media (www.springer.com)

Preface

ScicosLab (`http://www.scicoslab.org`) is a free open-source software package for scientific computation. ScicosLab includes a fork of Scilab, based on Scilab 4, the modeling and simulation tool Scicos and a number of other toolboxes.

Scilab is an interpreted language specifically developed for matrix based numerical computations. It includes hundreds of general purpose and specialized functions for numerical computation, organized in libraries called toolboxes that cover such areas as simulation, optimization, systems and control, and signal processing. These functions reduce considerably the burden of programming for scientific applications.

One important ScicosLab toolbox is Scicos. Scicos (`http://www.scicos.org`) provides a block-diagram graphical editor for the construction and simulation of dynamical systems. Scilab/Scicos is the only open-source alternative to commercial packages for dynamical system modeling and simulation packages such as MATLAB/Simulink and MATRIXx/SystemBuild. Widely used at universities and engineering schools, Scilab/Scicos has also gained ground in industrial environments. ScicosLab is developed and maintained by research groups, in particular, at INRIA[1] and ENPC.[2]

ScicosLab includes full Scilab and Scicos user's manuals, which are available with search capabilities in a help window. All commands, their syntax, and simple illustrative examples are given. While very useful in finding out the details of a particular command, these manuals do not provide a tutorial on the philosophy of either Scilab or Scicos. Nor do they address how to use several of these commands together in the solution of a technical problem.

The objective of this book is to provide a tutorial for the use of Scilab/Scicos with a special emphasis on modeling and simulation tools. While it will provide useful information to experienced users, it is designed to be accessible to beginning users from a variety of disciplines. Students [52] and academic and industrial scientists and engineers should find it useful. The discussion includes some information on modeling and simulation in order to assist the reader in deciding which simulation tools might be most useful to them. Every software environment has its special features, some would say quirks, that experienced users automatically take into account but often prove confusing to beginning users. We have tried to point these out where appropriate.

The book is divided into two parts. The first part concerns Scilab and includes a tutorial covering the language features, the data structures and specialized functions for doing graphics, importing and exporting data, interfacing with external routines, etc. It

---

[1] Institut National de Recherche en Informatique et en Automatique
[2] Ecole Nationale des Ponts et des Chaussées

also covers in detail Scilab numerical solvers for ODEs (ordinary differential equations) and DAE's (differential-algebraic equations). Even though the emphasis is placed on modeling and simulation applications, this part provides a global view of the product.

The second part is dedicated to modeling and simulation of dynamical systems in Scicos. Scicos provides a block-diagram editor for constructing models. This type of modeling tool is widely used in industry because it provides a means for constructing modular and reusable models. This part contains a detailed description of the editor and its usage, which is illustrated through numerous examples. It also covers advanced subjects such as constructing new blocks and batch simulation. Code generation and debugging are other topics covered. Finally, a new extension of Scicos is discussed. This extension allows the use of components described by the Modelica (`http://www.modelica.org`) language.

There have been several previous books written about Scilab. Most of them have been in French [20, 3, 2, 1, 29] and dealt with earlier versions of Scilab, as in [16]. The current book, like the earlier [18], is unique in a number of ways. They focus on simulation and modeling, put a major emphasis on Scicos, and discuss Scicos in depth. However, unlike [18] which was written before ScicosLab, *Modeling and Simulation in Scilab/Scicos with ScicosLab 4.4* is the first to deal with the new ScicosLab 4.4 version and the latest version of Scicos, Scicos 4.4.

The source of all the examples presented in this book can be downloaded from `http://www.scicos.org`.

Finally, a large number of people have supported us in many ways. We would especially like to thank our wives, Gail Campbell and Homa Nikoukhah, and our parents, Aline and René Chancelier, for their support in this and everything else we do.

Steve Campbell
Jean-Philippe Chancelier
Ramine Nikoukhah

Contents

Part I

Scilab

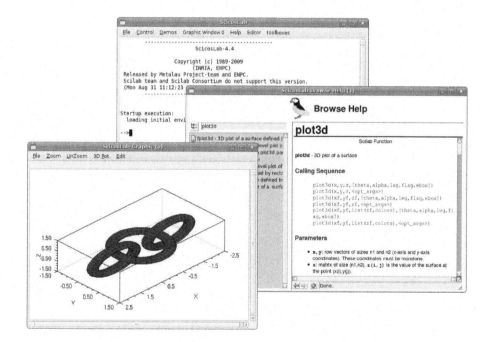

1

General Information

1.1 What Is Scilab?

There exist two categories of general scientific software: computer algebra systems that perform symbolic computations, and general purpose numerical systems performing numerical computations and designed specifically for scientific applications. The best-known examples in the first category are Maple, Mathematica, Maxima, Axiom, and MuPad. The second category represents a larger market dominated by MATLAB. Scilab belongs to this second category.

In this book we consider exclusively the Scilab/Scicos distributed in the ScicosLab package, which is free open-source software. This package is maintained by the research team at INRIA and ENPC that originally developed Scilab and is used for distributing the latest version of Scicos developed at the Metalau project at INRIA.

ScicosLab runs, and is available in binary format, for the main available platforms: Unix/Linux workstations (the main software development is performed on Linux workstations), Windows, and MacOSX. MacOSX users can also install Scilab using fink. Compiling Scilab from the source code is also possible and is fairly straightforward.

Scilab is an interpreted language with dynamically typed objects. Scilab can be used as a scripting language to test algorithms or to perform numerical computations. But it is also a programming language, and the standard Scilab library contains around 2000 Scilab coded functions. The Scilab syntax is simple, and the use of matrices, which are the fundamental object of scientific calculus, is facilitated through specific functions and operators. These matrices can be of different types including real, complex, string, polynomial, and rational. Scilab programs are thus quite compact and most of the time are smaller than their equivalents in C, C++, or Java.

Scilab is mainly dedicated to scientific computing, and it provides easy access to large numerical libraries from such areas as linear algebra, numerical integration, and optimization. It is also simple to extend the Scilab environment. One can easily import new functionalities from external libraries into Scilab by using static or dynamic links. It is also possible to define new data types using Scilab structures and to overload standard operators for new data types. Numerous toolboxes that add specialized functions to Scilab are available on the official site.

Scilab also provides many visualization functionalities including 2D, 3D, contour and parametric plots, and animation. Graphics can be exported in various formats such as Gif, Postscript, Postscript-Latex, and Xfig. In addition to Scilab's user interface functions, the Scilab Tcl/Tk interface can be used to develop sophisticated GUI's (Graphical user interfaces).

S.L. Campbell et al., *Modeling and Simulation in Scilab/Scicos with ScicosLab 4.4*, DOI 10.1007/978-1-4419-5527-2_1, © Springer Science+Business Media, LLC 2010

ScicosLab is a large software package containing approximately 13,000 files, more than 400,000 lines of source code (in C and Fortran), 70,000 lines of Scilab code (specialized libraries), 80,000 lines of online help, and 18,000 lines of configuration files. These files include

- Elementary functions of scientific calculation;
- Linear algebra, sparse matrices;
- Polynomials and rational functions;
- Classic and robust control, LMI optimization;
- Nonlinear methods (optimization, ODE and DAE solvers, Scicos, which is a hybrid dynamic systems modeler and simulator);
- Signal processing;
- Random sampling and statistics;
- Graphs (algorithms, visualization);
- Graphics, animation;
- Parallelism using PVM;
- MATLAB-to-Scilab translator;
- A large number of contributions for various areas.

1.2 How to Start?

1.2.1 Installation

ScicosLab is available for downloading at http://www.scicoslab.org. The procedure for installing ScicosLab depends on the operating system (Windows, MacOSX, Linux, or Unix), and information can be found on the website. The user has the choice between installing the binary version (if one is available for the host system) and compiling the source version. To compile the source version, the host system must be equipped with appropriate C and Fortran compilers. For the Windows operating systems, the binary version is obtained by cross compilation under Linux. Note that C and Fortran compilers are already installed on most Linux platforms, and for other Unix systems, if native compilers are not available, freely available GNU compilers can be used. On computers running MacOSX there is also the option of installing ScicosLab using fink.

In this book, SCI designates the directory in which ScicosLab is installed. We use the Unix notation for specifying path names. For example, SCI/routines/machine.h is the file machine.h in the subdirectory routines. Under the Windows operating system, all the " / " should be replaced with " \ ".

Even though there are no restrictions on its use, ScicosLab is copyrighted; see the file license.txt (English) or licence.txt (French) included in the package for restrictions on the redistribution of modified versions of the software.

1.2.2 First Steps

Running ScicosLab opens up a command window; see Figure 1.1. The look of this window may differ depending on the window manager.

The ScicosLab command window provides an interactive scilab shell where the user is invited to enter a command at a prompt (-->). The command must be validated with a carriage return, after which Scilab executes the command and returns control to the user by displaying a new prompt.

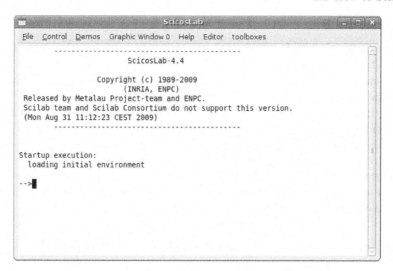

Figure 1.1. Scilab's main window.

The best way to start exploring ScicosLab is to run the demos. This can be done by clicking on the **Demos** button at the top of the command window (under the Windows operating system, the **Demos** button is in the "?" menu). The demos are chosen to present typical uses of the software and some of its specialized toolboxes.

For each demo, the user can see the corresponding Scilab source code, which shows that the data types used in Scilab are, for the most part, vectors and matrices. Their usage, very close to the usual matrix notation, results in compact and readable code. This, and the fact that there is no need for type declaration, compilation, or memory allocation, makes Scilab a lot easier to use than low-level languages such as C and Fortran. Just as for any other interpreted language, however, there is a price to be paid in terms of efficiency. This could become a factor in some applications.

1.2.3 Line Editor

Limited editing facilities are available in the command shell. Besides the usual cut and paste operations, line editing can be done using control characters as is done in `emacs`: `Ctrl-b` (pressing `b` while holding the `Ctrl` key down) for moving the cursor back by one character, `Ctrl-f` for moving it forward, `Ctrl-a` to place the cursor at the beginning of the command line, and `Ctrl-e` for placing it at the end. Also `Ctrl-k` erases the part of the command line between the current position of the cursor and the end of the line and saves it in a buffer, and `Ctrl-y` inserts the content of the buffer at the current position of the cursor. Previously entered commands can be searched using up and down arrows or equivalently with `Ctrl-p` and `Ctrl-n`.

Under the Unix and Linux operating systems, an additional feature is provided for searching and recalling a previously entered command by using the `Ctrl-r` (`reverse-i-search`) command and typing the string to be searched for.

All the commands entered in Scilab are automatically saved in a file called `scilab.hist` in the user's home directory.

1.2.4 Documentation

ScicosLab has a comprehensive online help facility, which can be consulted through the commands **help** and **apropos**. To consult the manual page corresponding to a Scilab function, the command **help** followed by the name of the function can be used. This opens up a browser window displaying the manual page in question. The manual page contains a detailed description of the function and a number of examples of its usage. The examples can be cut and pasted into ScicosLab's command window to be executed. The browser, which can also be accessed by clicking on the **help** button at the top of the command window, contains a list of all the functions classified by theme in different chapters (see Fig. 1.2). The manual page of a function can then be obtained by clicking on its name.

To obtain a list of Scilab functions corresponding to a keyword, the command **apropos** followed by the keyword should be used.

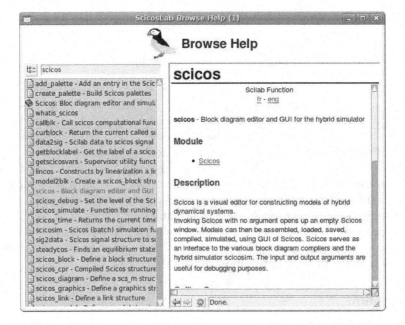

Figure 1.2. Help browser window.

The demos are also a good source of inspiration. They present simple examples of Scilab programming situations frequently encountered by users. The graphics demos, for example, give an overall picture of what can be done in Scilab as far as visualization is concerned. The demos' source codes often provide a good starting point for developing complex applications.

The demos provide examples for doing graphics, signal processing, systems control, Scilab simulation (in particular, the famous bicycle example), Scicos simulation, and a lot more.

1.3 Typical Usage

A typical ScicosLab user spends most of his time going back and forth between the Scilab shell and a text editor. For the most part, Scilab programs contain very few lines of

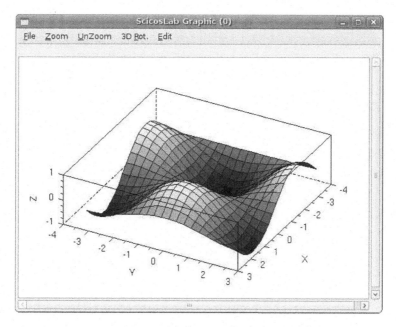

Figure 1.3. Graphic window.

instructions, thanks to the powerful data types and primitives available in Scilab. These programs can be written using a built-in editor, which is activated through the `Editor` menu button or using the user's favorite text editor (for example `vi` or `emacs`, for which there exists a Scilab mode, under Unix, and `notepad` or `wordpad` under Windows), then loaded into Scilab. This can be done through the built-in editor menus or using Scilab functions `getf` and `exec` (discussed in Section 2.2), or through the menu `File operations`.

1.4 ScicosLab on the Web

The latest release of ScicosLab, its documentation, and many third-party contributions (toolboxes) can be found on the official ScicosLab home page at

`http://www.scicoslab.org`

There is also a newsgroup dedicated to Scilab:

`comp.soft-sys.math.scilab`

A specific site devoted to teaching with Scilab is also available at

`http://cermics.enpc.fr/scilab/`

Finally, the email address `support@scicoslab.org` can be used for contacting the developers concerning contributions or simply for asking questions.

2

Introduction to Scilab

The Scilab language was initially devoted to matrix operations, and scientific and engineering applications were its main target. But over time, it has considerably evolved, and currently the Scilab language includes powerful operators for manipulating a large class of basic objects.

In Scilab, objects are never declared or allocated explicitly; they have a dynamic type and their sizes can dynamically change according to applied operators or functions. The following Scilab session illustrates the creation of an object, its dynamic change, and its deletion. Note that the function `typeof` returns the type of the object, `isdef` tests the presence of the object, and `clear` destroys it and frees the corresponding memory.

```
⊢⟶ a= rand(2,3);                              ← 2x3 random matrix a is created
⊢⟶ typeof(a)                   ← type of a, constant stands for real or complex matrix
   ans  =

 constant

⊢⟶ a= [a, zeros(2,1)]                         ← size of a dynamically increases
   a  =

!   0.2113249    0.0002211    0.6653811    0. !
!   0.7560439    0.3303271    0.6283918    0. !

⊢⟶ a= 'scilab'; typeof(a)                     ← redefining a through reassignment
   ans  =

 string

⊢⟶ exists('a')                                ← checks if a exists
   ans  =

   1.

⊢⟶ clear('a'); exists('a')        ← removing a explicitly from current environ-
                                   ment and verifying that a is gone
   ans  =

   0.
```

S.L. Campbell et al., *Modeling and Simulation in Scilab/Scicos with ScicosLab 4.4*, DOI 10.1007/978-1-4419-5527-2_2, © Springer Science+Business Media, LLC 2010

Available memory for a Scilab session can be controlled through the use of the `stacksize` command. It is of course bounded by available dynamic heap memory (allocated memory through the use of the `malloc` function). Using Scilab commands `who` and `whos`, one can get the name and size of dynamic objects present in the current session:

```
⟼ stacksize()                        ← available memory (number of double) and maxi-
  ans  =                               mum number of variables


!   1000000.    6185. !

⟼ A=rand(1000,1000);
                 !--error   17
rand: stack size exceeded (Use stacksize function to increase it)

⟼ stacksize(6000000);                        ← increasing Scilab available memory
⟼ A=rand(1000,1000);                 ← A is created, note that an expression end-
                                       ing with ; is evaluated without display
⟼ whos('-name','A');                         ← get info on variable A
Name                  Type         Size         Bytes

A                     constant     1000 by 1000  8000016
```

By default, numbers in Scilab are coded as double-precision floats. The accuracy of double-precision computation is machine-dependent and can be obtained through the predefined Scilab variable `%eps`, called the machine precision. `%eps` is the largest double-precision number for which `1+(%eps)/2` is indistinguishable from 1.

The set of numbers in Scilab can be extended by adding `%inf` (infinity) and `%nan` (not a number). This is done with the Scilab function `ieee`:

```
⟼ ieee()                        ← get the default floating-point exception mode
  ans  =


  0.

⟼ 1/0                           ← 1/0 raises an error in the default mode
   !--error   27
division by zero...

⟼ ieee(2)                       ← using the standard ieee floating-point exception mode
⟼ 1/0                                        ← 1/0 evaluates now to  + ∞
  ans  =


  Inf

⟼ 0/0                                        ← no error is produced
  ans  =


  Nan
```

Scilab expressions are evaluated by the `scilab` interpreter. The interpretor has access to a large set of functions. Some of these functions are Scilab-coded functions (we shall see later that Scilab is a programming language), some are hard-coded functions (`C` or `Fortran` coded). This ability of Scilab to include `C` or `Fortran` coded functions means that a large amount of existing scientific and engineering software can be used from within Scilab.

Scilab is a rich and powerful programming environment because of this large number of available functions. The users can also enrich the basic Scilab environment with their own Scilab or hard-coded functions, and moreover this can be done dynamically (hard-coded functions can be dynamically loaded in a running Scilab session); we shall see how this can be done later. Thus, toolboxes can be developed to enrich and adapt Scilab to specific needs or areas of application.

2.1 Scilab Objects

Scilab provides users with a large variety of basic objects starting with numbers and character strings up to more sophisticated objects such as booleans, polynomials, and structures. A Scilab object is a basic object or a set of basic objects arranged in a vector, a matrix, a hypermatrix, or a structure (list).

As already mentioned, Scilab is devoted to scientific computing, and the basic object is a two-dimensional matrix with floating-point double-precision number entries. In fact, a real scalar is nothing but a 1×1 matrix. The next Scilab session illustrates this type of object. Note that some numerical constants are predefined in Scilab. Their corresponding variable names start with %. In particular, π is `%pi`, and e, the base of the natural log, is `%e`. Note also that `a:b:c` gives numbers starting from `a` to `c` spaced `b` units apart.

```
⊢─→ a=1:0.6:3                          ← a is now a scalar matrix (double)
  a  =

!   1.    1.6    2.2    2.8  !

⊢─→ b=[%e,%pi]                ← b is the 1x2 row vector filled with predefined values (e, π)
  b  =

!   2.7182818    3.1415927  !
```

New types have been added as Scilab has evolved, but the matrix aspect has always been kept. String matrices, boolean matrices, sparse matrices, integer matrices (`int8`, `int16`, `int32`), polynomial matrices, and rational matrices are now available in the standard Scilab environment. Complex structures called lists (`list`, `tlist`, and `mlist`) are also available. Note also that functions in Scilab are considered as objects as well.

```
⊢─→ a= "Scilab"                                ← a 1x1 string matrix
  a  =

  Scilab

⊢─→ b=rand(2,2)                                ← a matrix
  b  =

!   0.2113249    0.0002211  !
!   0.7560439    0.3303271  !
```

```
┌──→ b= b>= 0.5                                              ← a boolean matrix
 b  =


! F F !
! T F !

┌──→ L=list(a,b)                                             ← a list
 L  =


        L(1)

 Scilab

        L(2)

! F F !
! T F !

┌──→ A.x = 32;A.y = %t                   ← a structure implemented using mlist
 A  =

   x: 32
   y: %t

┌──→ a= spec(rand(3,3))              ← eigenvalues: a vector of complex numbers
 a  =


!    1.8925237                !
!    0.1887123 + 0.0535764i !
!    0.1887123 - 0.0535764i !
```

It is possible to define new types in Scilab in the sense that it is possible to define objects whose dynamic type (the value returned by `typeof`) is user-defined. Scilab operators, for example the `+` and `*` operators, can be overloaded for these dynamically defined types. The new types are defined and implemented with `tlist` and `mlist` primitive types (see Section 2.1.6).

2.1.1 Matrix Construction and Manipulation

As already pointed out, one of the goals of Scilab is to give access to matrix operations for any kind of matrix types. In this section we highlight general functions and operators that are common to all matrix types.

A matrix in Scilab refers to one- or two-dimensional arrays, which are internally stored as a one-dimensional array (two-dimensional arrays are stored in column order). It is therefore always possible to access matrix entries with one or two indices. Vectors and scalars are stored as matrices.

Multidimensional matrices can also be used in Scilab. They are called hypermatrices.

Elementary construction operators, which are overloaded for all matrix types, are the row concatenation operator ";" and the column concatenation operator ",". These two operators perform the concatenation operation when used in a matrix context, that is, when they appear between "[" and "]". All the associated entries must be of the same type. Note that in the same matrix context a white space means the same thing as "," and a line feed means the same thing as ";". However, this equivalence can lead to confusion when, for example, a space appears inside an entry, as illustrated in the following:

⟼ A=[1,2,3 +5]   ← here A=[1,2,3,+5] with a unary +
  A =

! 1.   2.   3.   5. !

⟼ A=[1,2,3 *5]   ← here A=[1,2,3*5] with a binary *
  A =

! 1.   2.   15. !

⟼ A=[A,0; 1,2,3,4]
  A =

! 1.   2.   15.   0. !
! 1.   2.   3.   4. !

' and .'	transpose (conjugate or not)
diag	(m,n) matrix with given diagonal (or diagonal extraction)
eye	(m,n) matrix with one on the main diagonal
grand	(m,n) random matrix
int	integer part of a matrix
linspace or " : "	linearly spaced vector
logspace	logarithmically spaced vector
matrix	reshape an (m,n) matrix (m*n is kept constant)
ones	(m,n) matrix consisting of ones
rand	(m,n) random matrix (uniform or Gaussian)
zeros	(m,n) matrix consisting of zeros
.*.	Kronecker operator

Table 2.1. A set of functions for creating matrices.

Table 2.1 describes frequently used matrix functions that can be used to create special matrices. These matrix functions are illustrated in the following examples:

⟼ A= [eye(2,1), 3*ones(2,3); linspace(3,9,4); zeros(1,4)]
  A =

! 1.   3.   3.   3. !
! 0.   3.   3.   3. !
! 3.   5.   7.   9. !
! 0.   0.   0.   0. !

⟼ d=diag(A)'   ← main diagonal of A as a row matrix
  d =

```
!   1.    3.    7.    0. !
```

→ B=diag(d)   ← builds a diagonal matrix
 B   =

```
!   1.    0.    0.    0. !
!   0.    3.    0.    0. !
!   0.    0.    7.    0. !
!   0.    0.    0.    0. !
```

→ C=matrix(d,2,2)   ← reshape vector d
 C   =

```
!   1.    7. !
!   3.    0. !
```

The majority of Scilab functions are implemented in such a way that they accept matrix arguments. Most of the time this is implemented by applying mathematical functions elementwise. For example, the exponential function `exp` applied to a matrix returns the elementwise exponential and differs from the important matrix exponential function `expm`.

→ A=rand(2,2);   ← a random matrix (uniform law)
→ B=exp(A)   ← elementwise exponential
 B   =

```
!    1.2353136    1.0002212 !
!    2.1298336    1.3914232 !
```

→ B=expm(A)   ← square matrix exponential
 B   =

```
!    1.2354211    0.0002901 !
!    0.9918216    1.391535  !
```

Extraction, Insertion, and Deletion

To specify a set of matrix entries for the matrix A we use the syntax A(B) or A(B,C), where B and C are numeric or boolean matrices that are used as indices. The interpretation of A(B) and A(B,C) depends on whether it is on the left- or right-hand side of an assignment = if an assignment is present.

If we have A(B) or A(B,C) on the left, and the right-hand side evaluates to a nonnull matrix, then an assignment operation is performed. In that case the left member expression stands for a submatrix description whose entries are replaced by the ones of the right matrix entries. Of course, the right and left submatrices must have compatible sizes, that is, they must have the same size, or the right-hand-side matrix must be a scalar.

If the right-hand-side expression evaluates to an empty matrix, then the operation is a deletion operation. Entries of the left-hand-side matrix expression are deleted. Assignment or deletion can change dynamically the size of the left-hand-side matrix.

```
⟼ clear A;
⟼ A(2,4) = 1                                    ← assigns 1 to (2,4) entry of A
 A  =

    0.    0.    0.    0.
    0.    0.    0.    1.

⟼ A([1,2],[1,2])=int(5*rand(2,2))              ← assignment for changing a submatrix of A
 A  =

    1.    0.    0.    0.
    3.    1.    0.    1.

⟼ A([1,2],[1,3])=[]                            ← submatrix deletion
 A  =

    0.    0.
    1.    1.

⟼ A(:,1)= 8                                    ← ":" stands for all indices, here all the rows of A
 A  =

    8.    0.
    8.    1.

⟼ A(:,$)=[]                                    ← deletion, "$" stands for the last index
 A  =

    8.
    8.

⟼ A(:,$+1)=[4;5]                               ← adding a new column through assignment
 A  =

    8.    4.
    8.    5.
```

When an expression A(B) or A(B,C) is not the left member of an assignment, then it stands for a submatrix extraction and its evaluation builds a new matrix.

```
⟼ A = int(10*rand(3,7));                                  ← int integer part
⟼ B=A([1,3],$-1:$)                             ← extracts a submatrix  using row 1 and 3
                                                 and the two last columns of A
 B  =

!   2.    8. !
!   2.    3. !
```

When B and C are boolean vectors and A is a numerical matrix, A(B) and A(B,C) specify a submatrix of matrix A where the indices of the submatrix are those for which B and C take the boolean value T. We shall see more on that in the section on boolean matrices.

Elementary Matrix Operations

Table 2.2 contains common operators for matrix types and in particular numerical matrices. If an operator is not defined for a given type, it can be overloaded. The operators in the table are listed in increasing order of precedence. In each row of the table, operators share the same precedence with left associativity except for the power operators, which are right associative.

Operators whose name starts with the dot symbol "." generally stand for term-by-term (elementwise) operations. For example, C=A.*B is the term-by-term multiplication of matrix A and matrix B, which is the matrix C with entries C(i,j)=A(i,j)*B(i,j).

|	logical or
&	logical and
~	logical not
==, >=, <=, >, < ,<>, ~=	comparison operators
+,-	binary addition and subtraction
+,-	unary addition and subtraction
.*, ./, .\, .*., ./., .\., *, /, /., \.	"multiplications" and "divisions"
^,** , .^ , .**	power
', .'	transpose

Table 2.2. Operator precedence.

Note that these operators include many important equation-solving operations such as least squared solutions. The following calculations illustrate the use of several of these operators.

```
⊢→ A=(1:3)' * ones(1,3)          ← transpose ' and usual matrix product *
   A   =
```

```
!   1.    1.    1. !
!   2.    2.    2. !
!   3.    3.    3. !
```

```
⊢→ A.* A'                         ← multiplication tables: using a term-by-term product
   ans   =
```

```
!   1.    2.    3. !
!   2.    4.    6. !
!   3.    6.    9. !
```

```
⊢→ t=(1:3)';m=size(t,'r');n=3;
⊢→ A=(t*ones(1,n+1)).^(ones(m,1)*[0:n])   ← term-by-term exponentiation to build
                                             a Vandermonde matrix  A(i,j) = t_i^{j-1}
   A   =
```

```
!   1.    1.    1.    1.  !
!   1.    2.    4.    8.  !
!   1.    3.    9.    27. !
```

```
⊢→ A=eye(2,2).*.[1,2;3,4]         ← Kronecker product
   A   =
```

```
! 1.   2.   0.   0. !
! 3.   4.   0.   0. !
! 0.   0.   1.   2. !
! 0.   0.   3.   4. !
```

```
⟼ A=[1,2;3,4];b=[5;6];
⟼ x = A \ b ; norm(A*x -b)                    ← \ for solving a linear system Ax=b.
  ans  =

    0.
```

```
⟼ A1=[A,zeros(A)]; x = A1 \ b                 ← underdetermined system: a solution with
  x  =                                          minimum norm is returned


! - 4.  !
!   4.5 !
!   0.  !
!   0.  !
```

```
⟼ A1=[A;A]; x = A1\ [b;7;8]                   ← overdetermined system: a least squared
  x  =                                          solution is returned


! - 5.  !
!   5.5 !
```

2.1.2 Strings

Strings in Scilab are delimited by either single or double quotes " ' " or " " ", which
are equivalent. If one of these two characters is to be inserted in a string, it has to be
preceded by a delimiter, which is again a single or double quote. Basic operations on
strings are the concatenation (operator " + ") and the function length, which gives the
string length. As expected, a matrix whose entries are strings can be built in Scilab, and
the two previous operations extend to string matrix arguments as do the usual row and
column concatenation operators. A string is just a 1x1 string matrix whose type is denoted
by string (as returned by typeof).

```
⟼ S="a string with a quote character <<"'>> "
  S =

  a string with a quote character <<'>>

⟼ S='a long string 0...                      ← ... used to continue on next line
⟼     using continuation '
  S =

  a long string     using continuation

⟼ S=['A','string';'2x2','matrix']                         ← a string matrix
  S =
```

```
!A     string  !
!              !
!2x2   matrix  !
```

⟼ length(S)   ← length of each string of S in a matrix
  ans =

```
!  1.    6. !
!  3.    6. !
```

ascii	conversion from string to ascii values
execstr	send a string to the Scilab interpreter
grep	search for occurences of a string in a string matrix
part	substring extraction
msscanf	scans input from a string according to a format
msprintf	builds a string by output according to a format
strindex	finds occurrences of strings in a string
string	converts data to string
stripblanks	remove leading and trailing white (blank) characters
strsubst	string substitution in a string matrix
tokens	string tokenizer
strcat	concatenate string matrix entries
length	length of string matrix entries

Table 2.3. Some string matrix functions.

String matrix utility functions are listed in Table 2.3. The next session shows how some of them can be used.

⟼ A=rand(2,8,'n');   ← Normal law
⟼ A=sign(A);   ← just keep the signs
⟼ A=string(A)   ← convert to string matrix
  A =

```
!-1  1  1  1  1   1   -1  1  !
!                           !
!1   1  1  1  -1  -1  -1  1  !
```

⟼ A=strsubst(A,'1','+');   ← string substitution
⟼ A=strsubst(A,'-+','-')   ← string substitution
  A =

```
!-  +  +  +  +  +  -  +  !
!                       !
!+  +  +  +  -  -  -  +  !
```

The command execstr can be used to evaluate a string as a Scilab expression. The given string is evaluated using values from the current context. Thus, string matrices can be used to build Scilab expressions, which then can be evaluated as if they were entered interactively in Scilab. An important extra argument `'errcatch'` can be given to the `execstr` function in order to supress the Scilab standard error mechanism while the string is evaluated.

```
⟼ name ='x'; n=3; val=[45,67,34];
⟼ str = name +string(1:n)+'=val(' +string(1:n)+');'          ← a string vector
 str  =

!x1=val(1);  x2=val(2);  x3=val(3);   !

⟼ execstr(str);                                               ← Scilab evaluation of str
⟼ [x1,x2,x3]                                                  ← x1, x2 and x3 are now defined
 ans  =

!  45.   67.    34. !
```

2.1.3 Boolean Matrices

A boolean variable can take only the two values "true" T and "false" F. Two predefined Scilab variables `%t` and `%f` respectively evaluate to T and F and can be used to build boolean matrices, for example through the use of concatenation operations.

Comparison operators (" `==` ", " `>` ", " `>=` ", " `<=` " and "`~=`") also give boolean matrices as the result when their arguments are matrices or one matrix and one scalar. Logical operators such as " `&` " (and) , " `|` " (or), and "`~`" (not) can be used as expected with boolean matrix arguments. The logical function **and** (resp. **or**) takes as argument a single boolean matrix and returns the logical and (resp. or) of the matrix entries.

Boolean matrices are used in Scilab in conjunction with conditional expressions such as the **if** and **while** conditions, which will be described later.

```
⟼ true=%t;                                                    ← define boolean variable
⟼ if true  then disp("Hello"), end                            ← conditional display
 Hello
```

The following Scilab session shows simple instructions involving booleans. We see in particular that even though matrix booleans are not coded as numbers, they can be used in numerical computations:

```
⟼ ~(1>=2)
 ans  =

 T

⟼ %t&%t
 ans  =

 T

⟼ x=-10:0.1:10;
```

```
⟼ y=((x>=0).*exp(-x))+((x<0).*exp(x));        ← automatic boolean-to-scalar conversion
⟼ y=bool2s([%t,%f])                           ← explicit boolean-to-scalar conversion
 y  =
```

```
!  1.    0. !
```

We have mentioned previously that submatrix extraction can be done with boolean vectors. This is illustrated in the following session, where we also introduce the function **find**, which returns the indices of the true entries of a boolean matrix.

```
⟼ A = int(10*rand(1,7))
 A  =
```

```
!  2.    7.    0.    3.    6.    6.    8. !
```

```
⟼ A( A>= 3) = 0                               ← indices are given by a boolean matrix A>=5
 A  =
```

```
!  2.    0.    0.    0.    0.    0.    0. !
```

```
⟼ I=find(A== 0)                               ← indices of A entries equal to 0
 I  =
```

```
!  2.    3.    4.    5.    6.    7. !
```

2.1.4 Polynomial Matrices

Polynomials are Scilab objects. Most operations available for constant matrices are also available for polynomial matrices. A polynomial can be defined using the Scilab function **poly**. It can be defined based either on its roots or its coefficients, as can be seen from the following Scilab session.

```
⟼ p=poly([1 3],'s')                           ← polynomial defined by its roots
 p  =

              2
   3 - 4s + s
```

```
⟼ q=poly([1 2],'s','c')                       ← polynomial defined by its coefficients
 q  =

   1 + 2s
```

Note that the `'s'` argument in **poly** specifies the character to be used for displaying the formal parameter of the polynomial. At initialization, the variable %s is defined to be the polynomial s.

Polynomials can be added together, multiplied, concatenated to form matrices, etc., provided they use the same formal parameter.

⟼ p+q+1   ← polynomial/constant addition
  ans =

                   2
    5 - 2s + s

⟼ [p*q,1]   ← matrix polynomial
  ans =

!   2   3   !
!   3 + 2s - 7s + 2s   1   !

A polynomial can even be divided by a polynomial, and a polynomial matrix inverted, but the result, in general, is not a polynomial. It is a rational function, which is another Scilab object constructed using a list object, as we shall see later.

2.1.5 Sparse Matrices

Many areas of application lead to problems involving large matrices. But very often, large matrices involved in applications have many zero entries. Thus representations of matrices where matrix storage is reduced by coding only the nonzero matrix entries have been developed and are known as sparse matrix representations. Different coding schemes exist for performing this compression task and some schemes are tailored to specific applications. In Scilab, the sparse coding of an (m,n) matrix is as follows. A first array of size m contains pointers to row descriptors. Each row descriptor is coded with two arrays. The first array contains the column indices of the nonnull row entries, and the second array contains the associated nonzero values.

Some problems that could not be stored in Scilab memory in full form can be accessible through the use of sparse storage. As one can imagine, the reduced storage leads to more complex and slower operations on sparse matrices than on full matrices. The basic function used for building sparse matrices in Scilab is the **sparse** function. Note that only numerical or boolean matrices can be stored as sparse matrices.

⟼ A=sprand(100,100,0.1);   ← a (100,100) sparse matrix with 10% of
                                        nonzero values
⟼ whos('-type','sparse')   ← check used memory

Name	Type	Size	Bytes
A	sparse	100 by 100	13360

⟼ B=full(A);
⟼ whos('-name','B');   ← used memory for the same matrix with a
                                        full implementation

Name	Type	Size	Bytes
B	constant	100 by 100	80016

⟼ timer();inv(B);timer()   ← CPU time for full matrix inversion
  ans =

    0.01

```
⊢→ timer();inv(A);timer()                         ← CPU time for sparse matrix inversion
  ans  =

    0.28
```

While useful in illustrating the commands, this example was only 100×100, which is not considered today to be a very large matrix. Also, this matrix was random, and many sparse matrices, such as those that arise by the discretization of PDEs, are not random. They have stronger structural properties that can be exploited.

2.1.6 Lists

Scilab lists are built with the `list`, `tlist`, and `mlist` functions. These three functions do not exactly build lists, but they can be considered to be structure builders in the sense that they are used to aggregate under a unique variable name a set of objects of different types. They are implemented as an array of variable-size objects that is not a list implementation. A type corresponds to each builder function, and they are recursive types (a list element can be a list).

- If the `list` constructor is used, then the stored objects are accessed by an index giving their position in the list.
- If the `tlist` constructor is used, then the built object has a new dynamic type and stored objects can be accessed through names. Note also that the fields are dynamic, which means that new fields can be dynamically added or removed from an occurrence of a `tlist`. A `tlist` remains a list, and access to stored objects through indices is also possible.
- The `mlist` constructor is a slight variation of the `tlist` constructor. The only difference is that the predefined access to stored objects through indices is no longer effective (it is, however, possible using the `getfield` and `setfield` functions). Also, extraction and insertion operators can be overloaded for `mlist` objects. This means that it is possible to give a meaning to multi-indices extraction or insertion operations. `hypermat` objects that implement multidimensional arrays are implemented using `mlist` in Scilab.

Note that many Scilab objects are implemented as `tlist` and `mlist`, and from the user point of view this is not important. For example, suppose that you want to define a variable with an extendable number of fields. This is done very easily through the use of the "." operator:

```
⊢→ x.color = 4;
⊢→ x.value = rand(1,3);
⊢→ x.name = 'foo';
⊢→ x                                              ← x is a tlist of type struct (st)
  ans  =

    color: 4
    value: [0.2113249,0.7560439,0.0002211]
    name: "foo"
```

Among Scilab objects coded this way, it is important to mention the rational matrices and the linear state-space systems, which play important roles in modeling and analysis of linear systems.

The following session illustrates the use of rational matrices in Scilab (recall that %s is by default the polynomial s). Note that all the elementary operations are overloaded for rational matrices.

```
⟼ r=1/%s                                    ← defining a rational number
  r  =

      1
      -
      s

⟼ a=[1,r;1,1]                               ← rational matrix construction
  a  =

  !   1       1   !
  !   -       -   !
  !   1       s   !
  !               !
  !   1       1   !
  !   -       -   !
  !   1       1   !

⟼ b=inv(a)                                  ← matrix inversion
  b  =

  !     s        - 1     !
  !   -----      -----   !
  ! - 1 + s    - 1 + s   !
  !                      !
  !    - s         s     !
  !   -----      -----   !
  ! - 1 + s    - 1 + s   !

⟼ b.num                                     ← numerator field
  ans  =

  !   s      - 1   !
  !               !
  ! - s       s   !

⟼ b.den                                     ← denominator field
  ans  =

  ! - 1 + s    - 1 + s   !
  !                      !
  ! - 1 + s    - 1 + s   !
```

A linear state-space system is characterized in terms of four matrices, A, B, C, and D; we will describe and use these systems later for modeling linear systems. The Scilab function ssrand defines a random system with given input, output, and state sizes.

```
⟼ sys=ssrand(1,1,2)                                          ← defines a linear system
  sys  =

        sys(1)    (state-space system:)

!lss  A  B  C  D  X0  dt  !

        sys(2) = A matrix =

! - 0.7616491    1.4739762 !
!   0.6755537    1.1443051 !

        sys(3) = B matrix =

!   0.8529775 !
!   0.4529708 !

        sys(4) = C matrix =

!   0.7223316    1.9273333 !

        sys(5) = D matrix =

    0.

        sys(6) = X0 (initial state) =

!   0. !
!   0. !

        sys(7) = Time domain =

    c

⟼ sys.A                                                      ← extract the A matrix
  ans  =

! - 0.7616491    1.4739762 !
!   0.6755537    1.1443051 !
```

Common operations on lists are illustrated in the following example. Note that one can define an entry starting somewhere other than entry 1, but then the list will create early places for entries. This is analogous to Scilab automatically setting up a 3×5 matrix when the instruction A(3,5)=6.7 is entered. Except now, instead of extra zero entries, an entry called Undefined fills out the earlier positions in the list.

```
⟼ L=list()                                                  ← an empty list
  L  =

    ()
```

```
⟼ L(2) = testmatrix('magi',3)
```
← list assignment using L(i)=val. Note that
L(1) does not exist

```
 L  =

        L(1)

    Undefined

        L(2)

!   8.    1.    6. !
!   3.    5.    7. !
!   4.    9.    2. !

⟼ L(0) = 34;
⟼ L($) = 'X'
 L  =

        L(1)

    34.

        L(2)

    Undefined

        L(3)

 X
```
← add an element at the begining of list
← replace last element

```
⟼ L($+1) = 'Y'
 L  =

        L(1)

    34.

        L(2)

    Undefined

        L(3)

 X

        L(4)

 Y
```
← add an element at end of list

```
⟼ [a,b]=L([1,3])
 b  =

 X
 a  =
```
← extraction, we can extract several argument in one call

```
34.
```

⊢⟶ L(2)=null();   ← deletion of the second list element

2.1.7 Functions

We have used many Scilab functions in previous examples through function calls without precisely defining what a function is. A function is known through its calling syntax. For example, the calling syntax for the sine function `sin` is the same as that for the hyperbolic sine function `sinh`.

There is a small difference between `sin` and `sinh`. `sin` is a hard-coded function (hard-coded functions are sometimes called primitives) and `sinh` is a Scilab-coded function (they are sometimes called macros in Scilab). Thus, `sin` and `sinh` do not have the same type. The answer to `typeof` will be `fptr` for `sin` and `function` for `sinh`. But both `sin` and `sinh` are Scilab objects. They have a type, and can be used as variables and function arguments.

A Scilab-coded function can be defined interactively using the keywords `function` and `endfunction`, loaded from a scilab script using `exec` or `getf`, or saved and loaded in binary mode using `save` and `load`. In addition, a library of functions can be defined and loaded in Scilab. At startup, a number of function libraries are loaded, providing a rich set of functions available to the user. The source code of a Scilab-coded functions can be examined using the `fun2string` function.

We will discuss the construction of new functions in Scilab later: in Section 2.2.3 we go over Scilab-coded functions, and in Section 2.5, hard-coded functions are covered. Here, we look at an elementary example to give a feeling for the syntax.

⊢⟶ **function y=foo(x,g) ; y=g(x); endfunction**   ← a function
⊢⟶ **typeof(foo)**   ← a function is a Scilab object
   ans =

   function

⊢⟶ foo(%pi,sin)   ← functions can be used as function arguments (primitive `sin`)
   ans =

    1.225D-16

⊢⟶ foo(%pi,sinh)== sinh(%pi)   ← the same with the macro `sinh`
   ans =

   T

⊢⟶ v=rand(1,10);
⊢⟶ foo(3,v)   ← function call and matrix extraction have the same syntax.
   ans =

    0.0002211

As with other variables, functions can be removed or masked by assignment. For example, if a session starts with the command `sin=4`, then the `sin` variable is 4 and no longer the `sin` primitive. However, primitives and scilab functions available through libraries (for example those provided by Scilab default libraries) are just hidden by this mechanism. If `sin` is cleared by the command `clear sin`, then the `sin` primitive will return to the current environment.

```
⟼ sin = [1,3];
```

```
⟼ sin(2)                      ← the sin primitive is hidden by the defined variable
  ans  =
```

```
  3.
```

```
⟼ clear sin                                   ← clear the sin variable
```

```
⟼ sin(2)                               ← the sin primitive is found and called
  ans  =
```

```
  0.9092974
```

2.2 Scilab Programming

A Scilab program is a set of instructions to be executed in a specific order. These instructions can be typed one by one at Scilab's prompt, but they can also be placed in an ASCII file (using for example Scilab's built-in editor) and executed with the Scilab command `exec`. Such a file is called a *script* and may contain function definitions. By convention, Scilab script file names are terminated with the suffix `.sce`, but if the script contains only function definitions, then the `.sci` suffix is used. In that case the command `getf` can be used instead of `exec` to load the functions.

The file extensions `.sci` and `.sce` are just conventions, since Scilab functions `exec` and `getf` do not check file name suffixes. However, in most editors with a Scilab mode, the mode activation is controlled by the file-name suffix. Under the Windows operating system, a file can be drag-and-dropped into Scilab. In this case, the file extensions are used to select appropriate actions such as loading with `getf` and executing with `exec`.

It is also possible to execute a script when launching Scilab from a shell window (Unix or MSDOS), using the calling option `scilab -f <script-file-name>`. If the script is terminated by the instruction `quit`, then Scilab will quit at the end of the script execution. This provides for batch mode execution in Scilab.

When Scilab is launched, a specific script file called `scilab.star` is executed. This file contains, in particular, instructions for initializing a number of Scilab variables and loading various libraries. Then, if there exists a file called `.scilab` or `scilab.ini` in the user's home directory (the home directory is defined by the environment variable `HOME`), Scilab executes it as a script. Finally, Scilab looks in the directory where Scilab is launched and executes the file `.scilab` or `scilab.ini`, if it exists.

Iteration and branching instructions are a fundamental part of programming, so we begin by looking at these instructions in Scilab.

2.2.1 Branching

Branching instructions are used to make block execution of code depend on boolean conditions. The simplest form of branching instruction in Scilab takes the following form:

```
if <condition> then <instructions> end
```

The block of instructions `<instructions>` will be executed if the condition (`<condition>`) evaluates to a boolean `T`. Since Scilab is a matrix language, the condition evaluation can be a boolean matrix or a scalar matrix. If it is a matrix, the condition is considered as a `T` boolean only if all the matrix entries are true boolean values or if all the scalar matrix entries are nonnull. Thus, implicitly `<condition>` is evaluated as an `and(<condition>)`. The following session generates the logarithm of a random 3×3 matrix and checks whether it is real.

```
⟼ A=log(rand(3,3)) ;
⟼ if imag(A)==0  then disp('A is a real matrix ');  end
 A is a real matrix
```

The following syntax is used for the two branch form:

```
if <condition> then <instructions> else <instructions> end
```

The first block is executed when `<condition>` evaluates to `T`; otherwise, the second block is evaluated.

```
⟼ if imag(A)==0  then
⟼    disp('A is a real matrix ');
 A is a real matrix
⟼ else
⟼    disp('A is complex');
⟼ end
```

A multiple branch version is also available. Successive `else` statements are coded using the `elseif` keyword. Note that there also exists a multiple branch execution instruction named `select`, which can be used when execution control depends on a set of predefined values:

```
select <expr> ,
  case <expr1> then <instructions>
  case <expr2> then <instructions>
  ...
  else <instructions>
end
```

In executing `select`, the value of the `<expr>` statement is successively compared to the values of `<expr1>`, `<expr2>`, ... As soon as both evaluated expressions evaluate to equal values (`==` operator), the execution branches to the associated block. If no equal case is detected, then the `else` block, if present, will be executed.

A special case of branching is obtained with the try-catch instruction:

```
try <try-statements> catch <catch-statements> end
```

This instruction enables control of errors in block execution. The `<try-statements>` block is executed and if an error occurs during its execution, the execution branches to the `<catch-statements>`.

```
⟼ x= [1,2];
⟼ try
⟼   y = [4;5] + x                          ← inconsistent addition
⟼ catch
⟼   lasterror();                           ← clear recorded error message
⟼   y = [4;5]' + x                         ← try a new addition
  y  =

      5.    7.
⟼ end
```

2.2.2 Iterations

Two iterative control structures exist in Scilab. They are the `for` loop and the `while` iterator. For execution efficiency, one should carefully check to see wether iterative control structures can be replaced by matrix operations, which usually execute faster.

The `for` loop syntax is as follows:

```
for <name>=<expr>
   <instructions>
end
```

The `<expr>` instruction is evaluated once. Then, the inner block will be iteratively executed, and at each iteration the `<name>` variable will take a new value. If the `<expr>` instruction evaluation gives a matrix, then the number of iterations is given by the number of columns, and the loop variable `<name>` will take as value the successive columns of the matrix. If the `<expr>` instruction evaluation gives a list, then the number of iterations is given by the list length, and the loop variable `<name>` will take as value the successive values of the list elements.

A `break` statement can be present in the `<instructions>` block of the `for` loop. If the `break` instruction is reached during execution, then the block execution stops and execution continues beyond the end of the `for` loop.

The following Scilab session illustrates several aspects of using `for`, including the differences in CPU time when the alternative of using matrix operations exists. This session uses the function `pmodulo(n,i)`, which returns the remainder upon dividing n by i, n mod i. The session also uses the `timer` function to find the CPU time for a computation.

```
⟼ n=89;
⟼ isprime=%t;
⟼ for i=2:(n-1)                           ← iterate on integers from 2 to n-1
⟼    if pmodulo(n,i)==0  then isprime=%f;  break; end
⟼ end

⟼ isprime                                 ← checks result stored in variable isprime
  ans  =
```

T

```
⊢⟶ n=16778;                              ← now illustrate difference in CPU time

⊢⟶ timer();                                            ← start timer
⊢⟶ res=[];                                ← want all the divisors of n
⊢⟶ for i=2:(n-1)
⊢⟶    if pmodulo(n,i)==0 then
⊢⟶       res=[res,i];               ← size of vector res increases at each iteration
⊢⟶    end
⊢⟶ end
⊢⟶ t1=timer();                    ← CPU time elapsed from last call to timer

⊢⟶ res                                      ← all the divisors of n
  ans  =

!  2.    8389. !

⊢⟶ v=2:(n-1);            ← speeding up computation using matrix computation
⊢⟶ timer();
⊢⟶ I=find(pmodulo(n,v)==0);              ← indices of divisors using find
⊢⟶ res = v(I)                                ← all the divisors of n
 res  =

!  2.    8389. !

⊢⟶ t2=timer();
⊢⟶ [t1,t2]                              ← the CPU time of each example
  ans  =

!  0.83    0.01 !
```

The `while` iterator is especially useful when it is not known ahead of time how many iterations are to be performed. It has the following syntax:

```
while <condition>
    <instructions>
end
```

The `<instructions>` block statements are executed while the `<condition>` evaluates to boolean T. As in the `if` statement, the condition evaluation can be a boolean matrix or a numerical matrix. If it is a matrix, then `<condition>` is considered as a T boolean only if all the matrix entries are true boolean values or if all the matrix entries are nonzero. If `<condition>` always evaluates to T, then a `break` statement will be needed to stop the `<while>` loop. The next three examples use the fact that large enough real numbers are considered to be infinity `%inf`, which can be compared or added to finite real numbers.

```
⊢⟶ x=1; while exp(x)<>%inf ; x=x+1; end      ← a simple while with a scalar condi-
                                               tion
⊢⟶ [exp(x-1),exp(x)]==%inf                     ← checks result
  ans  =

! F T !
```

⟼ x=[1:3]; while exp(x)<>%inf ; x=x+1; end   ← a simple while with matrix condition

⟼ exp(x)==%inf   ← one entry is %inf
  ans =

! F F T !

⟼ x=1;
⟼ while %t   ← infinite loop, need a **break** to quit the **while**
⟼   if exp(x)== %inf then   ← quit the loop when $\exp(x)$ equals ∞
⟼   break;
⟼   end
⟼   x=x+1;
⟼ end
⟼ [exp(x-1),exp(x)]==%inf   ← checks result
  ans =

! F T !

2.2.3 Scilab Functions

It is possible to define new functions in Scilab. What distinguishes a function from a script is that a function has a local environment that communicates with the outside through input and output arguments.

A function is defined using the keyword `function` followed by its calling syntax, a set of Scilab instructions, and the keyword `endfunction`. More precisely, the following form of syntax is used to define a function:

```
function [<name1>,<name2>,...]=<name-of-function>(<arg1>,<arg2>,...)
   <instructions>
endfunction
```

When defining a function we have to give the function name (`<name-of-function>`), the list of calling arguments (`<arg1>,<arg2>,...`), and the list of variables that are used to return values (`<name1>,<name2>,...`). Note that a function can return more than one value.

Calling a function is done as follows:

```
   <name-of-function>(<expr1>,<expr2>,...)
```

or

```
   [<v1>,<v2>,...,<vp>]=<name-of-function>(<expr1>,<expr2>,...)
```

In the first case the returned value is the value of the first returned argument by the function evaluation. In the second case the `p` first returned values of the function evaluation will be copied in the `p` variables whose names are given by [`<v1>,<v2>,...,<vp>`].

When a function is called, the expressions given as arguments are first evaluated and their values are passed to the function evaluation. Thus, Scilab uses a calling by value mechanism for all argument types. However, if an argument of a calling function instruction is a variable name and if that variable is not changed in the function evaluation body, then the variable will not be copied during the function call.

In the evaluation of the body of a function, variable search is performed first in the local environment of the function evaluation and then in the calling environments. Thus a variable that is not a calling variable and not locally defined in the function can still have a value if it is defined in a calling environment. However, the variable in the calling environment cannot be changed. If an assignment statement involving this variable is used, then a local copy will be created.

The function body evaluation normally stops when all the <instructions> are executed or when the flow of instructions reaches a return statement. When function evaluation stops, the flow of instruction returns to the caller. The returned values <name1>, <name2>, ... have the values that they had when the function body evaluation stopped.

The following session gives two examples of defining a function and evaluating it. The second function defines the factorial function.

```
⟼  function B=f(A)                    ← a typical function with one input and one output
⟼    B=string(sign(A));
⟼    B=strsubst(strsubst(B,'1','+'),'-+','-');
⟼  endfunction
⟼ f(rand(2,5,'n') )
  ans  =

!-  +  +  +  +  !
!               !
!+  +  +  +  -  !
```

```
⟼  function y=fact(x)                 ← a recursive function can be defined: fact calls fact
⟼    if x <= 1  then y=x;  else y=x*fact(x-1); end
⟼  endfunction

⟼ fact(4)
  ans  =

  24.
```

This next script defines several functions and illustrates the rules for searching variable values when a function is evaluated. This is important but quite technical and can be omitted for first-time readers.

```
⟼  function y=f(x); y=2*x; endfunction

⟼ x=90;
⟼ f()                                 ← argument x is not given but exists in the
                                        calling environment. Thus, the calling se-
                                        quence works
  ans  =

  180.
```

```
⟼ f(5,7)                              ← error: two many given arguments
       !--error    58
incorrect number of arguments in  function call...
arguments are  :
  x
```

⟼ [a,b]=f(5)   ← error: asking for too many returned arguments
        !--error   59
incorrect # of outputs in the function
arguments are :
 y

⟼ function y=f(x); z=x; endfunction   ← a new definition of f

⟼ y=89;
⟼ z=67;
⟼ w=f(x)   ← y is not computed inside f but it has a
                                             value in the calling environment. This is
  w =   the returned value

    89.

⟼ z   ← z was not modified by execution of f
  ans =

    67.

⟼ function y=f(); y=x; endfunction

⟼ x=5;
⟼ y=f()   ← x is not locally defined: the value of x in the calling environment is used
 y =

    5.

⟼ function y=f(); x= 2*x; y=x; endfunction

⟼ y=f()   ← a local variable x is created
 y =

    10.

⟼ x=[56,67];
⟼ function y=f(); x(1)= 5; y=x   endfunction

⟼ y=f()   ← again a local variable x is created leading to a scalar y
 y =

    5.

As already pointed out, a function that is specified with n calling arguments and p returned values can be used in a calling statement with fewer than n calling values, and the number of requested returned arguments can be less than p.

During execution of a function body the actual number of given arguments and the number of requested outputs are available through the use of the function **argn**. A statement like [lhs,rhs]=argn() in a function body will return in **lhs** (resp. **rhs**) the number of requested output arguments (resp. the number of given input arguments). The function writer can take advantage of this for enabling optional argument passing, as shown in the following example. This example also shows that the function `error` can be used to produce an error statement in Scilab.

```
⟼  function [u,v]=f(x,y)
⟼     [lhs,rhs]=argn()
⟼     if rhs <= 0  then error('at least one argument must be given'); end
⟼     if rhs <= 1  then y=2; end
⟼     if lhs == 2  then
⟼        u=x; v=y;
⟼     else
⟼        u=x+y;
⟼     end
⟼  endfunction

⟼  [u,v]=f(4)
 v  =

    2.
 u  =

    4.
```

If the function defined has n calling arguments, then you do not normally have the right to call a function with more than n calling arguments. There is one exception, which is described now. If the last argument of a function definition with n arguments is named `varargin`, then the function can be called with more than n arguments. Inside the function body, the `varargin` variable will be a Scilab list that contain as elements the values of the arguments from the nth one to the last one.

```
⟼  function [l]=f(x,varargin); l = varargin;  endfunction
⟼ f(0,1,2)                                        ← varagin is set to list (1,2)
 ans  =

    ans(1)

 1.

    ans(2)

 2.
```

The same mechanism also exists for the output through the keyword `varargout`, as illustrated in the following session.

```
⟼  function [varargout]=f()
⟼     varargout=list(1,2,3)
⟼  endfunction

⟼  [a,b]=f()                                       ← f can be called with more than 1 output
 b  =
```

```
        2.
   a  =

        1.
```

Changing a variable in a calling environment, as a side effect of a function call, can sometimes be useful. This is achieved in Scilab through the use of global variables. The `global` keyword can be used to declare a global variable. Once a variable is declared global in an environment its value can be changed from within that environment. For example, a `global('x')` statement must be present in the body definition of a function if the function is to be used to modify this global variable x.

```
⟼ global a;              ← a is now a global variable, its default value is []
⟼ isglobal(a)                                    ← just checks ...
  ans  =

   T

⟼ function f(); global('a'); a=int(10*rand(1,4));  endfunction
⟼ f()                    ← calling f, which contains a global declaration of a
⟼ a                                        ← calling f has changed a
  ans  =

!  2.    7.    0.    3. !
```

2.2.4 Debugging Programs

When programming in any language, it is nearly impossible to write large programs without errors. Programming in Scilab is not an exception to this rule. A set of Scilab utilities helps users detect and correct bugs in their code. In particular, one can interrupt the execution of Scilab code using the command **pause** (menu `Stop` of the `Control` menu in the main menu) or using the interruption key code `Ctrl-C` .

When entering a pause, the Scilab prompt changes, thereby signaling to the user that the interpreter has entered a pause sequence, and shows the pause level. When a function is called, its code is executed in a local environment. The same situation occurs when entering a pause. The current execution flow is stopped and the interpreter is called in a new local environment. One has to consider that the environments are stacked and the current environment is at the top of the stack. When entering a function call or a pause, a new environment is pushed to the top of the stack. When code is executed in the environment that is at the top of the stack, it has read access to variables from the other stacked environments and read-write access to the global environment. When a pause is entered through the use of `Ctrl-C`, execution stops at the current execution position. Using `whereami` (or `where`), it is possible to detect where the pause has occurred.

It is, however, easier to explicitly insert a **pause** command in Scilab code or in a function in order to precisely control where a pause is needed to detect wrong execution of code. This can be done in an editor by explicitly inserting **pause** statements.

```
⟼ a=34;
```

⟼ **function** y=f(x); pause; a = %pi; pause; y=g(x);   **endfunction**
⟼ **function** y=g(x); b = %e ; y=sin(x); pause;   **endfunction**

⟼ f(5)
-1->a   ← a pause in **f**. We can check variables from the stacked environments
  ans =

    34.

-1->b= 56;   ← here we create a variable in the pause environment

-1->resume   ← quit the first pause and stop at the second one still in **f**

-1->a   ← the value of **a** in the local environment of **f**
  ans =

    3.1415927

-1->resume   ← quit the pause and stop in the next one in **g**

-1->exists('b','local')   ← the value of **b** in **g** environment
  ans =

    0.

-1->resume   ← go on
  ans =

  - 0.9589243

⟼ a=g(4);   ← stop in **g**
-1->[y]=resume(456);   ← quit the pause environment and copy a
                                        variable in function **g** environment

⟼ a   ← the returned **g** value is the **y** value that was set by the **resume** function
  ans =

    456.

Leaving the pause mode and ending the debugging is done with the `abort` command.
   Changing function code by inserting `pause` commands implies that you have to reload
the function code in Scilab each time you change the code. The function `setbpt` can be
used to do the same task but without changing function code.
   In the local environment of the pause, one can check the status of variables, and the
current execution can be resumed with the command `resume`. Execution flow can also be
interrupted using `abort` (resp. `quit`). Execution is then stopped and all the environments
are popped from the stack in the case of an `abort` command. Just the current environment
is popped in the case of a `quit` command. (Here, getting something off the stack is called
a pop and putting something on the stack is called a push.)
   Step-by-step execution of code or of a function can also be enabled through the use of
the `exec` command. When a function call is performed with `exec`, the function is executed
in the current environment and one has to provide values for the function calling arguments
in the current environment before calling `exec`.

⟼ **function** y=f1(x)
⟼   y=sin(x).*x
⟼   y = y+2,
⟼ **endfunction**

⟼ getf('stepf.sci','n');   ← load **f1** with **getf** option 'n'
⟼ x=int(10*rand(1,4));   ← argument of **f1**
⟼ x=90;
⟼ exec. (f1,7)   ← single-step execution **f1**
step-by-step mode: enter carriage return to proceed
>>
y=sin(x).*x
  y =

     80.4597
>>
y = y+2,
  y =

     82.4597
>>
return!

Note that by default, the source code of the function instructions is hidden when getf is used. In order to see the source code, one has to load the function into the Scilab environment with the getf command associated with the optional argument 'n'.

Finally, it should be noted that the built-in editor also provides some debugging facilities.

2.3 Input and Output Functions

The default for Scilab commands is to show the results in the command window unless they create a graphic, and then the default is to show the result in a graphics window. However, it is often desired to have the result saved in a file or printed on a printer. In this section we discuss how to control the input and output of different types of data.

2.3.1 Display of Variables

As already seen in previous examples, when a Scilab instruction is followed by a comma or a line feed, a display of the instruction evaluation result is provided. If instructions are ended by a semicolon or belong to the body of a function, then no display occurs.

Sometimes one wants values of variables inside of functions to be displayed or output to a file. Explicit display of the value of a variable, or of the result of an instruction evaluation, is achieved by the function disp or print. These two functions provide a display similar to the one obtained automatically in the command window. The first argument of the print function is used to control the output stream (disp uses the Scilab command window as output stream). One can use a file name given by a Scilab string or a file descriptor (returned by the file command). Alternatively one can use the predefined value %io(2), which stands for the standard output stream in the Scilab command window.

⟼ a=[%pi,4,%inf]   ← the result is displayed
  a =

!   3.1415927   4.   Inf !

⟼ print(%io(2),a)   ← using **print** to display a variable (same as **disp**)
  a =

!   3.1415927   4.   Inf !

⟼ print('a.txt',a)   ← **print** to file **a.txt**

print can be used to force the display of variables inside a function during its execution, as illustrated below:

⟼ **function** y=f(x) ; a = 2*%pi, y = x+a, **endfunction**
⟼ f(2);   ← a is not displayed
⟼ **function** y=g(x) ; a = 2*%pi, print(%io(2),a), y = x+a, **endfunction**
⟼ g(2);   ← force display of a using **print**
  a =

    6.2831853

The format used by **disp** and **print** for displaying real numbers can be controlled using the **format** function.

2.3.2 Formatted Input and Output

Several Scilab functions can used to perform formatted input-output. The functions **write** and **read** are based on **Fortran** formatted input and output. They redirect input and output to streams, which are obtained through the use of the function **file**. They can also use the standard Scilab input-output streams, which are given by the predefined variables %io(2) and %io(1).

We will not describe these functions in detail here although they can be quite useful for interfacing with Fortran software. We will focus here on Scilab functions that emulate standard C input and output functions. They give access to formatted (ASCII or binary) input or output on the standard input-output streams, on file streams, and on strings.

The function **mopen** is a Scilab implementation of the standard C function **fopen** with the following syntax:

```
[fd,err]=mopen(filename, mode, swap)
```

Here **filename** is a string that denotes a file name. The **mode** argument is also a string, whose length does not exceed three (its default value is 'rb') that describes the stream mode to be used. The input stream is obtained with ('r') and the output stream with ('w'). Append mode is obtained by adding an ('a') character, and a bidirectional stream for input and output is obtained with ('r+'). An added 'b' character will specify that streams are to be used in binary mode.

`mprintf`	formatted output on standard output stream
`mfprintf`	formatted output in a file stream
`msprintf`	formatted output in a string matrix
`mscanf`	formatted input on standard input stream
`mfscanf`	formatted input in a file stream
`msscanf`	formatted input in a string matrix
`fprintfMat`	scalar matrix output in a file
`fscanfMat`	read a scalar matrix from a file
`mgetl`	input a file as an `mx1` string matrix
`mputl`	output a string matrix in a file
`mopen`	open a file stream
`mclose`	close a file stream

Table 2.4. Input-output functions.

The `mclose` function can be used to close opened input or output streams, even the streams opened with the `file` command (the number of opened streams is system limited). It is possible to close all the opened streams using the instruction `mclose('all')`.

The formatted input and output functions are very similar to functions provided by the standard C input-output library. For example, the format specification follows the format specification of C functions. However, Scilab input-output formatted functions automatically take care of the matrix aspect of Scilab objects. When using a formatted output stream, the arguments of the conversion specifiers are searched in the rows of the Scilab arguments. When using a formatted input it is possible to specify the number of times the format has to be repeated, for example, up to the end of file. Thus it is possible using one read instruction to obtain a matrix, or a set of matrices, row after row:

```
⟼ mprintf('| %5d | %5.4f |\n', (1:2)',[%pi;%e])
```
← formatted output sequentially applied to rows

```
|     1 | 3.1416 |
|     2 | 2.7183 |
```

```
⟼  function table(A)
⟼    [m,n]=size(A);
⟼    format= "%5.0f |";                             ← format for each entry
⟼    format=strcat(format(ones(1,n)));              ← repeat the format n times
⟼    format1= part("-",ones(1,5+2));
⟼    format1=strcat(format1(ones(1,n)));                ← a string made of "-"
⟼    mprintf(format1+'\n');
⟼    mprintf('|'+format+'\n'+format1+'\n',A);      ← same format for each
                                                       row of A
⟼  endfunction
```

```
⟼ A=int(20*rand(2,5));
⟼ table(A)                          ← A dislayed as integer values in a table
```

```
-----------------------------------
|     4 |     0 |    13 |    16 |    17 |
-----------------------------------
|    15 |     6 |    12 |    13 |     1 |
-----------------------------------
```

```
⊢⟶ S=msprintf('%d %d \n',int(20*rand(3,2)));        ← using a string matrix as output (S)

⊢⟶ A= rand(4,3);
⊢⟶ fprintfMat('test',A);  B=fscanfMat('test');      ← read and write of a constant matrix
                                                      in an ascii file
⊢⟶ norm(B-A)                                         ← testing the write-read sequence
  ans =

     0.0000007

⊢⟶ fd=mopen('test','r');                             ← opens a file for reading
⊢⟶ L=mfscanf(-1,fd,"%f%f%f")                         ← scanning a file with the same format
                                                      on each row up to end of file
  L =

!   0.231224     0.3076090     0.3616360 !
!   0.216463     0.932962      0.292227  !
!   0.883389     0.214601      0.5664250 !
!   0.6525130    0.3126420     0.482647  !

⊢⟶ norm(L-A)          ← note that numerical precision depends on selected format directives
  ans =

     0.0000007

⊢⟶ mclose(fd);                                       ← close the opened file
```

The next example describes a formatted input where strings and scalars are mixed in a file that contains information on some Scicos files. The file `mfscanf.dat` is the following

```
An example of reading a file with  a separator of type '[ ]*,[ ]*'
---------------------------------------------------------------------
CLKIN_f.sci    ,   98  ,   16
CLKINV_f.sci   ,   91  ,   16
CLOCK_f.sci    ,  116  ,   16
CONST_f.sci    ,   50  ,    8
CURV_f.sci     ,   80  ,   12
```

We now illustrate how to use the function `mfscanf` to read out separately the numerical and the string information in this file.

```
⊢⟶ fd=mopen('mfscanf.dat','r');                      ← opens a file in read mode
⊢⟶ mgetl(fd,2);                                      ← bypassing the first two lines
⊢⟶ [n,a,b,c]=mfscanf(-1,fd,'%[^,],%*[, ]%d%*[, ]%d\ n');  ← formatted read

⊢⟶ n                                                 ← number of read arguments
  ans =

     3.

⊢⟶ stripblanks(a)'                  ← removing spaces and tranposing string matrix
  ans =

!CLKIN_f.sci  CLKINV_f.sci  CLOCK_f.sci  CONST_f.sci  CURV_f.sci  !
```

⟼ [b,c]   ← numerical values
  **ans = **

```
!   98.    16.  !
!   91.    16.  !
!  116.    16.  !
!   50.     8.  !
!   80.    12.  !
```

⟼ mclose(fd);   ← close the opened file stream

2.3.3 Input Output in Binary Mode

Scilab has its own binary internal format for saving Scilab objects and Scilab graphics in a machine-independent way. It is possible to save variables in a binary file from the current environment using the command **save**. A set of saved variables can be loaded in a running Scilab environment with the **load** command. The functions **save** and **load** use a machine-independent binary format. Note that the advantage of the binary mode is that **save** and **load** can be done without any loss of numerical accuracy. For saving and reloading graphics in a Scilab binary format, the functions **xsave**, **xload** can be used (note that these functions can also be accessed through menus).

It is also possible within Scilab to read and write any binary file through the use of a set of functions that will be described here. Note that as in the previous section, the functions described here need a stream as first argument. Input or output streams are created with the **mopen** function already described, which is to be used here with a binary flag ('rb' or 'wb').

Functions described in Table 2.5 can be used for binary input-output. The **mget** and **mput** use the mode (little or big "endian") that was specified by the **mopen** call, but it is always possible to disable this default behavior by using the appropriate optional argument. Full information on each command is found in the online help.

mget	binary input
mput	binary output
mgetstr	string input
mputstr	string output
mtell	returns current access position for an open stream
mseek	changes the access position for an open stream
meof	tests the end-of-file indicator

Table 2.5. Binary input-output functions.

In order to produce machine-independent binary input and output, the functions described in Table 2.5 always output data using the *little endian* format. It is possible to disable this default feature using the **swap** parameter. If **swap** is set to 0, then binary input and output is performed using the native format of the machine executing Scilab. This can be useful when binary files coming from other software are to be read into Scilab.

A set of functions for input and output of sound files is provided in the directory **SCI/macros/sound** in the Scilab main directory. They can be used as programming examples for the functions listed in Table 2.5.

We conclude this section with a small example dealing with the binary input and output of a constant matrix.

```
⟼ x=testmatrix('magic',4);                          ← a 4x4 test matrix
⟼ fd=mopen('save.dat','wb');              ← opens a file for writing in binary
                                             mode ('b' is useful on windows)
⟼ mput(length('x'),'i',fd)       ← writes an integer. The length of the string 'x'
⟼ mputstr('x',fd) ;                              ← writes a string 'x'
⟼ mput(size(x,'r'),'i',fd)          ← writes an integer (number of rows)
⟼ mput(size(x,'c'),'i',fd)          ← writes an integer (number of columns)
⟼ mput(x,'d',fd);                        ← writes an array of doubles
⟼ mclose(fd);                            ← closes the file output stream
⟼ clear x;


⟼ fd=mopen('save.dat','rb');        ← opens a file for reading in binary mode
⟼ ls=mget(1,'i',fd);                           ← reads an integer


⟼ name=mgetstr(ls,fd);           ← reads a string. The string length is given by ls
⟼ m=mget(1,'i',fd);n=mget(1,'i',fd);          ← reads two integers m and n
⟼ data=mget(m*n,'d',fd);                ← reads m*n doubles in a vector
⟼ code= msprintf('%s=matrix(data,%d,%d);',name,m,n);
⟼ execstr(code);            ← reshape the matrix and give it the correct name (i.e, x)
⟼ mclose(fd) ;                                  ← close the file
⟼ and(x == testmatrix('magic',4))                    ← test
  ans  =

  T
```

2.3.4 Accessing the Host System

As already noted, formatted and binary inputs and outputs use streams to access or create files. Streams are created and opened through the command `mopen` and closed using the `mclose` (or through the `file` command for Fortran-compatible streams). File streams are opened given a file pathname, which is coded as a Scilab string and can be absolute or relative. The preferred syntax for file names is the Posix-style syntax (which works on Windows operating systems as well), i.e., with a slash as directory separator. If necessary, the `pathconvert` function can be used for pathname translation. Note that a relative pathname is always considered relative to the Scilab current directory. The name of this current directory is returned by the function `getcwd` or `pwd` and can be changed using the `chdir` or `cd` function. Given a pathname, it is possible to access the information about a file using the `fileinfo` function. For example, `fileinfo('foo')<>[]` will evaluate to %t if file 'foo' exists. Finally, the list of files in a given directory can be obtained as a vector string using the commands `ls`, `dir`, and `listfiles`.

It is possible obtain information through global Scilab variables about the running Scilab program, in particular on the operating system it was built for and the embedded programs that were used at compilation. For example, the boolean variable `%tk` can be used to check wether the current Scilab version is built with interaction with Tcl/Tk (See also `%gtk` for Gtk support and `%pvm` for parallel virtual machine support). `MSDOS` is a boolean variable used to check wether the underlying operating system is Windows or a Unix-like system.

Running a shell script from Scilab is done using `unix` and related commands `unix_g`, `unix_s` (equivalently `host`), `unix_w`, and `unix_x`. These functions differ from each other in the way they return the result of the command: in the current Scilab window (`unix` and `unix_w`), in a graphics widget (`unix_x`), in a Scilab variable (`unix_g`), or without returned result (`unix_s`). Note that contrary to what their names suggest, these commands work on Unix, Mac, and Windows operating systems.

It is sometimes useful to use the `MSDOS` variable before using the `unix` command since the syntax of shell commands differs between Unix and Windows. This is illustrated in the following script:

```
⟼ %tk                                    ← checks wether we have Tcl/Tk support
  ans  =

  T

⟼ MSDOS                                  ← is it a Windows version of Scilab ?
  ans  =

  F

⟼ if ~MSDOS  then
⟼    os=unix_g('uname');
⟼ else
⟼    os='windows";
⟼ end
⟼ os                                     ← Operating system name ?
  ans  =

  Linux

⟼ if MSDOS  then
⟼    l=unix_g('dir sc*.sci');            ← list directory contents
⟼ else
⟼    l=unix_g('ls sc*.sci');             ← list directory contents
⟼ end
⟼ l
  l  =

!scalarbase.sci  !
!                !
!scalarops.sci   !
!                !
!scalar.sci      !
```

2.3.5 Graphical User Interface

So far, we have described input-output interaction with Scilab through keyboard interaction in the Scilab graphics command window or through predefined menus. It is also possible to interact with Scilab through dialog windows and to create specific dialog windows. We will discuss how to do this here.

Scilab menus (in the command window and in the graphics windows) can be customized using Scilab commands. `addmenu`, `delmenu`, `setmenu`, or `unsetmenu` can be respectively used to add, delete, activate, or deactivate Scilab menus.   One can associate a menu handler to a specified menu through a Scilab function. For example `addmenu('foo')` (resp. `addmenu('foo',4)`) will place a menu button `foo` in the Scilab command window (resp. graphics window 4), which when activated will call Scilab function `foo` (resp. `foo_4`).

A set of predefined popup dialog windows are also available in Scilab. Some examples are given in Figure 2.1. The simplest one is `x_message`, which is used to open a popup message window. Others are `x_choose`, which is used to perform a selection in a predefined list; `x_dialog`, which opens a dialog to get a response from the user as a Scilab string; `x_choices`, which is used to select a set of items each one being chosen through a toggle list; `x_mdialog`, which is used for a set of `x_dialog`-like interactions or for acquiring a matrix; and `xgetfile`, which is a dialog used to get a file path.

```
⊢→ rep=x_choose(['S.L.Campbell','J.P.Chancelier','R.Nikoukhah'],'Authors');
⊢→ x_message(['Well, it''s time ';'to get something to eat !']);

⊢→ chateaux=['Château Pétrus','Château Margaux',...
⊢→                'Château Talbot','Château Belgrave'];
⊢→ l1=list('Wine',1,chateaux);
⊢→ l2=list('Cheese',2,['Rocquefort','Ossau','Tome de Savoie']);
⊢→ rep=x_choices('Wine and Cheese menu',list(l1,l2));
```

Figure 2.1. Examples of predefined popup dialog windows.

Mouse and key events in graphics windows are returned by `xclick` and `xgetmouse`, and an event handler can be set for a graphics window with the function `seteventhandler`.

It is possible to build more sophisticated graphics user interfaces by mixing Scilab and Tcl/Tk code, which can interact through data exchange. This is made possible through a set of Scilab functions that interact with Tcl/Tk:

- `TCL_GetVar`, `TCL_SetVar` are used to exchange numerical data or data coded through strings.
- `TCL_ExistVar` is used to check for the existence of a Tcl/Tk variable.
- `TCL_EvalStr` is used to send a Tcl expression contained in a Scilab string to the Tcl interpreter for evaluation.
- `TCL_EvalFile` is used to send to the Tcl interpreter a file name of a Tcl script to be executed.
- `ScilabEval` is a Tcl instruction that can be used in a Tcl script. The Tcl string argument given to `ScilabEval` is evaluated as a Scilab expression.

We will give a small example of Tcl/Tk window creation with two range controls that can be used to change two scalar parameters a and b in a surface plot.

The following script defines a function called `plot3d_tcl`. This function uses global variables to keep track of two parameters a and b. In `plot3d_tcl`, calls are made to `TCL_GetVar` to get the value of Tcl variables named a and b. Since `TCL_GetVar` returns a string, a call to `evstr` is made to evaluate the string. If a or b in the Scilab environment differs from the Tcl/Tk values, then the Scilab values are updated and a surface is drawn. Graphics will be discussed in more detail in Section 2.4. Here we just note that `plot3d` will draw a surface and that the drawing of the surface depends on the two Scilab parameters a and b. Thus, if a or b is changed at the Tcl/Tk level, then a call to `plot3d_tcl` will draw a new surface. Note that a flicker effect will be seen in the graphics window when redrawing is performed. We will see in Section 2.4 how to avoid this.

```
function []=plot3d_tcl()
  global('a','b');
  a_new=evstr(TCL_GetVar("a"));
  b_new=evstr(TCL_GetVar("b"));
  if a<>a_new | b<>b_new then
    a=a_new ; b = b_new ;
    t=linspace(0,2,200);
    xbasc();t=linspace(-b*%pi,b*%pi,20);plot3d(t,t,a*sin(t)'*cos(t));
  end
endfunction
```

This function can be checked now with the following, which plots the surface:

```
⟼ TCL_SetVar('a','3');                    ← setting a to 3 in Tcl/Tk
⟼ TCL_SetVar('b','2');                    ← setting b to 3 in Tcl/Tk
⟼ plot3d_tcl();
```

Now we want a dialog window with two range controls in order to change the values of a and b interactively. This can be done as described in the next Tcl/Tk script.

```
# tcl/tk script
toplevel .top
wm resizable .top 0 0;
```

```
set a 5;
set b 2;
#——————————-
wm title .top {tk test};
global ok wait but;
proc runScilab {bidon} {ScilabEval plot3d_tcl()};
#—– top label
frame .top.widtop
label .top.widlabel −text {Surface parameters}
pack .top.widlabel −in .top.widtop −side top −pady 5
#—- a range control for a
frame .top.wid_frame2
label .top.widlabel_a −text {a}
scale .top.widscale_a −variable a −width 8 −state normal −from 1 −to 10 \
        −resolution .01 −orient horizontal −width 8 −command {runScilab}
pack .top.widlabel_a −in .top.wid_frame2 −side left −anchor w −padx 5 \
        −expand true
pack .top.widscale_a −in .top.wid_frame2 −side left −anchor e −padx 5
#—– a range control for b
frame .top.wid_frame3
label .top.widlabel_b −text {b}
scale .top.widscale_b −variable b −width 8 −state normal −from 1 −to 3 \
        −resolution .01 −orient horizontal −width 8 −command {runScilab}
pack .top.widlabel_b −in .top.wid_frame3 −side left −anchor w −padx 5 \
        −expand true
pack .top.widscale_b −in .top.wid_frame3 −side left −anchor e −padx 5
#——————-
pack .top.widtop −in .top
pack .top.wid_frame2 −in .top
pack .top.wid_frame3 −in .top
#——————-
raise .top
```

This is a pure Tcl/Tk script, wich is mostly independent of Scilab except for the fact that we call the `ScilabEval` function when a range control is changed. As already mentioned, `ScilabEval` is a Tcl function that sends its argument as a Scilab expression to the Scilab interpreter. Here, as expected, the `ScilabEval` function will send to the Scilab interpreter the string 'plot3d_tcl()'. Thus a range control modification will lead to a call to the `plot3d_tcl` function.

In order to launch the Tcl/Tk script we use `TCL_EvalFile` at the Scilab level. A picture of the Tcl/Tk dialog is provided in Figure 2.2

↦ `TCL_EvalFile("range.tcl");`   ← running the Tcl/Tk script from Scilab

For a restricted set of Tcl/Tk dialogs predefined in Scilab and called uicontrols, the process we have just described can be simplified in the sense that the Tcl/Tk script can be replaced by a Scilab script using the `figure`, `uicontrol`, and `uimenu` functions. The range controls used in the previous example are part of this restricted set of Tcl/Tk dialogs. We will now give the same example of a surface drawing but now managed through uicontrols. The function `plot3d_uicontrol` is the function that is activated when a range control (also

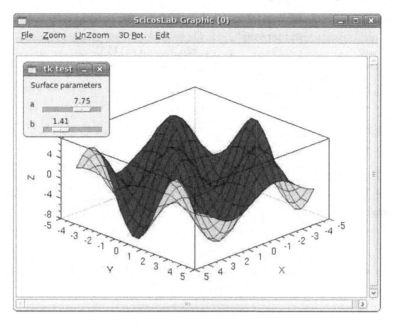

Figure 2.2. Example of a Tcl/Tk dialog.

called slider in uicontrol manual pages) is changed (this is called a callback), and the function myuidialog is used to build the Tcl/Tk window with the two sliders and the two text areas that display the sliders' values. The two functions share data through global Scilab variables. A picture of the Tcl/Tk dialog composed with uicontrols is provided in Figure 2.3

```
function plot3d_uicontrol()
  global('a_slider','a_text','b_slider','b_text','a','b','activate','n_fact');
  if activate == %t  then
    a=get(a_slider,'Value')/n_fact;
    b=get(b_slider,'Value')/n_fact;
    set(a_text,'String','a='+string(a))
    set(b_text,'String','b='+string(b))
    t=linspace(0,2,200);
    xbasc();t=linspace(-b*%pi,b*%pi,20);plot3d(t,t,a*sin(t)'*cos(t));
  end
endfunction

function myuidialog()
  global('a_slider','a_text','b_slider','b_text','a','b','activate','n_fact');
  n_fact=100;                        ← scaling because sliders work with integers
  activate=%f ;                      ← delay callback execution
  f = figure("Position",[50 50 200 60],"Unit", "pixel");
  a_text = uicontrol(f,"Position",[140 10 55 15],"Style","text",...
                  "String"    , "a=5")
  a_slider = uicontrol(f, "Position"  , [10 10 120 15],..
                    "Style"     , "slider",...
                    "Min"       , 1*n_fact,...
                    "Max"       , 10*n_fact,...
```

```
                     "Value"     , 5*n_fact,...
                     "SliderStep", [1 1],...
                     "callback"  , "plot3d_uicontrol()")
  b_text = uicontrol(f, "Position",[140 35 55 15],"Style","text",...
                     "String"    , "b=2")
  b_slider = uicontrol(f, "Position"   , [10 35 120 15],...
                     "Style"     , "slider",...
                     "Min"       , 1*n_fact,...
                     "Max"       , 3*n_fact,...
                     "Value"     , 2*n_fact,...
                     "SliderStep", [1 1],...
                     "callback"  , "plot3d_uicontrol()")
  activate=%t;                              ← now we accept callback calls
endfunction
```

Figure 2.3. Examples of uicontrol Tcl/Tk dialog.

2.4 Scilab Graphics

There are numerous Scilab graphics commands. A full discussion of them all would be a book by itself. Instead, we will first introduce some of the more basic graphics commands that are regularly used in simulation and modeling. Once users can generate the basic graphs they need, they start to want to do more sophisticated graphics. For these we will just give some general rules for performing graphics in the subsequent sections and give an example that illustrates a part of what is possible. Details can be found in the on-line manual.

2.4.1 Basic Graphing

One of the first things that someone interested in modeling and simulation wants to do is to generate plots. These plots can be graphs of functions, surfaces, curves, or data. Readers familiar with other matrix-based environments will have no problem with Scilab graphics. But readers coming from working with equation-based languages such as MAPLE and Mathematica will have to think a little bit differently. Scilab offers great control over how and where the graphics occur and how they look. But all this control can be confusing at first. In this section we show how to quickly and simply generate some of the more common types of graphics. The remaining subsections will describe how to modify the basic graphical output.

The `plot2d` command generates two-dimensional graphs. Given vectors

$$x = [x_1, \ldots, x_n] \text{ and } y = [y_1, \ldots, y_n],$$

plot2d(x,y) will plot the points $\{(x_1, y_1), \ldots, (x_n, y_n)\}$. There are several options for connecting the points, which can be specified using optional calling arguments. The default in plot2d is to connect them with a straight line. Some of the other most popular alternatives are already coded as plot2d2 , which assumes that the graph is a piecewise constant function, so it looks like a bar graph; plot2d3, which plots a vertical line up to each point; and plot2d4, which connects the plotted points with arrows.

By default, plot2d creates a graphics window called Scilab Graphic (0). This window can be cleared by the command clf or xbasc. Additional calls to plot2d with no clearing commands or redirection of output will result in all plots being on the same graph. Axes will automatically be adjusted.

It is of course possible to have many graphics windows at the time in a Scilab session. A window, which is described by an integer id i, can be selected for drawing by using the command xset('window',i). If the window does not exist, it will be created. Moreover, a graphics window can be partitioned into subwindows using the command subplot.

Later we will give some of the ways that the appearance of a plot can be altered with Scilab commands. But it is important to note that it is possible to make many of these changes using the Edit menu of the graphics window. Merely select Figure Properties. There is a menu bar at the top labeled Edit properties for:. This will give you menus for all the more important properties. The most commonly used ones are probably Axes and Polyline.

There are a number of ways to control the flow of Scilab output. Here we note that the Scilab graphics window has a file button. If clicked, it provides an easy way to export the graph into a number of formats including postscript. Note also that the graphics window has a zoom button, which allows you to select part of the graph and zoom in on it.

The fact that the graphics routines start with matrices sometimes creates problems. For example

```
⟼  tt=0:0.1:3;plot2d(tt,sin(tt));
```

will generate the graph of $\sin(t)$ on the interval $[0, 3]$. However,

```
⟼  function y=myf(t);y=sin(t)*sin(t);endfunction;
⟼  tt=0:0.1:3;plot2d(tt,f(tt))
```

will produce an error message and not the graph of $\sin^2(t)$. The problem is that the function myf has to be able to take the vector tt and evaluate it entry by entry. The following code will generate the graph of $\sin^2(t)$:

```
⟼  function y=myf(t);y=sin(t).*sin(t);endfunction
⟼  tt=0:0.1:3;plot2d(tt,f(tt))
```

Note that the function is defined using elementwise multiplication .* instead of *.

A number of graphs created using the basic versions of plot2d and some of its features can be found in Chapter 3

Plots of a surface $z = f(x, y)$ can be done in two ways. The scilab command plot3d works much like plot2d in that it has the syntax plot3d(x,y,z), where x, y, z are vectors. However, there is also the function fplot3d, which has the basic syntax fplot3d(x,y,f), where f is an external of the type y=f(x,y). Here x and y determine the grid on which f will be evaluated. The default for plot3d and fplot3d is for the graph to appear in a graphics window. Again the window can be exported into a postscript figure file by using one of the file options. Another useful feature is the 3D Rot. button of the graphics window. If this button is clicked, then one can click and hold on the figure, rotate using the mouse, and then click to release the figure. While the viewing point of the figure can also be adjusted by using optional calling arguments, it is usually easiest to find the best viewing angle by using the 3D Rot. option. Surface plots can be altered using the Edit menu of the graphics window just as 2D plots are.

2.4.2 Graphic Tour

We start with a number of scripts to make a tour of Scilab graphics functionalities. Two graphics styles coexist in Scilab today. But most high-level graphical functions are used in a similar fashion in both styles, so the examples provided in this first part work with both, as did the examples of the previous section. We shall later focus on the differences of the two styles and, in particular, present the object-oriented aspects of the new style.

Parametric Curve Plot in $\mathbb{R}^3$

The parametric plot of $(\sin(t), t\cos(t)/R, t/100)$ on $[-20\pi, 20\pi]$ using a uniform grid of 2000 points with $R = 20\pi$ is created by

```
⟼ t=linspace(-20*%pi,20*%pi,2000);
⟼ param3d1(sin(t),t.*cos(t)/max(t),t/100)          ← an ℝ³ curve (Figure~2.4)
```

The graph appears in Figure 2.4. The division by max(t) has the effect of scaling the second entry so that for any given interval it will have maximum value 1.

Plot a Surface in $\mathbb{R}^3$

The surface plot of $z = \sinh(x)\cos(y)$ on $-\pi \leq x \leq \pi$, $-\pi \leq y \leq \pi$ is created by

```
⟼ x=linspace(-%pi,%pi,40); y=linspace(-%pi,%pi,40);
⟼ plot3d(x,y,sinh(x')*cos(y)) ;                    ← an ℝ³ surface (Figure~2.5)
```

Note the transpose used on the x.

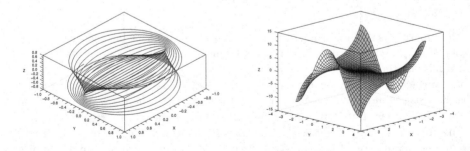

Figure 2.4. Parametric plot.   **Figure 2.5.** Surface plot in $\mathbb{R}^3$.

Plot a Vector Field in $\mathbb{R}^2$

The plot of the vector field for the differential equation

$$\dot{x}_1 = x_2 ,$$
$$\dot{x}_2 = -x_1(1 - x_1^2)x_2 ,$$

in the region $-1 \leq x \leq 1$, $-1 \leq y \leq 1$ is created by

```
⟼ function [xdot]=derpol(t,x);
⟼   xdot=[x(2);-x(1)+(1 - x(1)**2)*x(2)];
⟼ endfunction
⟼ xf= linspace(-1,1,10);yf= linspace(-1,1,10);
⟼ fchamp(derpol,0,xf,yf);                    ← an  ℝ² vector field (Figure~2.6)
```

Plot a Histogram

Generating a random vector with a normal distribution, plotting this vector as a histogram, and then plotting the normal distribution on the same graph is given by the following:

```
⟼ v=rand(1,2000,'n');
⟼ histplot([-6:0.4:6],v,[1],'015',' ',[-4,0,4,0.5],[2,2,2,1]);   ← a histogram
⟼ function [y]=f2(x); y=exp(-x.*x/2)/sqrt(2*%pi); endfunction;
⟼ x=-6:0.1:6;x=x';plot2d(x,f2(x),1);         ← superimposing an  ℝ² curve (Figure~2.7)
```

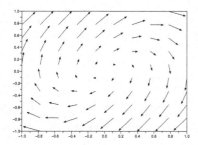
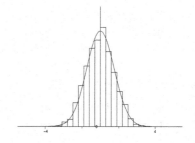

Figure 2.6. A vector field.   **Figure 2.7.** A histogram.

Polar Plot in $\mathbb{R}^2$

The plot in polar coordinates of $\rho = \sin(2\theta)\cos(2\theta)$ for $0 \leq \theta \leq 2\pi$ is created by the following commands:

```
⟼ theta=0:.01:2*%pi;rho=sin(2*theta).*cos(2*theta);
⟼ polarplot(theta,rho);                      ← an  ℝ² curve in polar coordinates (Figure~2.8)
```

Plot a Surface Given by Facets in $\mathbb{R}^3$

If a surface in $\mathbb{R}^3$ is parameterized by two variables, then by assigning a grid to these two variables we can create a surface composed of facets. This is illustrated below.

```
⟼ function [x,y,z]=f3(alpha,theta)
⟼   x=cos(alpha).*cos(theta);
⟼   y=cos(alpha).*sin(theta);
⟼   z=sinh(alpha);
⟼ endfunction
⟼ alphagrid=linspace(-%pi/2,%pi/2,40);thetagrid=linspace(0,2*%pi,20);
⟼ [x1,y1,z1]=eval3dp(f3, alphagrid, thetagrid);          ← building a set of facets
⟼ plot3d1(x1,y1,z1);              ← an  $\mathbb{R}^3$ surface composed by a set of facets (Figure~2.9)
```

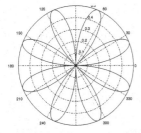

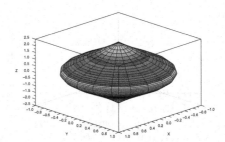

Figure 2.8. A polar plot.   **Figure 2.9.** A surface described by facets.

In the following examples, graphics objects are used explicitly. Thus these examples work only in the new graphics mode. Note, however, that the graphics functions `xrect` and `xpoly` can also be used in the old graphics mode.

Basic Graphics Functions: xrects

In the next example, each column of `rect` specifies a rectangle in terms of the coordinates of its upper left corner, width, and height. Then `xrects` plots these rectangles with a color determined by its second input variable.

```
⟼ xbasc()
⟼ a=get("current_axes");                    ← gets a handle on the newly created axes
⟼ a.title.font_size = 4;                         ← changes the title font size
⟼ a.data_bounds=[0,0;1,1];                ← sets the boundary values for x and z
⟼ xset('clipgrf');                    ← setting a clip zone in the graphic frame
⟼ n=20;
⟼ rects=[rand(1,n);rand(1,n);0.2*ones(1,n);0.2*ones(1,n)];
⟼ xrects(rects,rand(1,n)*20);                          ← a set of rectangles
⟼ xset('clipgrf');                    ← deactivates the clip zone
⟼ xtitle('A set of rectangles');                          ← (Figure~2.10)
```

Graphics handles and their properties will be discussed in the next section.

Basic Graphics Functions: `xpolys`

`xpolys` can be used to draw a set of line segments. The next example generates the end-points of several line segments and then calls `xpolys` to draw the line segments.

```
⟼ a=get("current_axes");                          ← gets a handle on the newly created axes
⟼ a.axes_visible="off";                              ← makes the axes invisible
⟼ a.data_bounds=[-2*cos(%pi/6),0;2*cos(%pi/6),3];   ← sets the boundary values for
                                                        x and z
⟼ a.box="off";
⟼ theta=[-%pi/6,%pi/2,%pi*7/6,-%pi/6];
⟼ T=[cos(theta);sin(theta)];
⟼ Ts=(1/(1+2*sin(%pi/6)))*rotate(T,%pi);
⟼ x=[T(1,:);Ts(1,:)]'; y=[T(2,:);Ts(2,:)]';
⟼ x=[x+cos(%pi/6),x-cos(%pi/6),x];
⟼ y=[y+sin(%pi/6),y+sin(%pi/6),y+1+2*sin(%pi/6)];
⟼ a.thickness=8;                                      ← sets line thickness
⟼ xpolys(x,y);                                        ← draws a set of polylines (Figure~2.11)
```

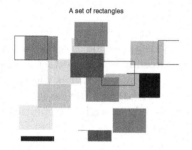

Figure 2.10. Drawing rectangles.

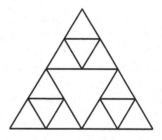

Figure 2.11. Drawing polygons.

More examples with animations are provided in Section 5.1.2 and Appendix B.2.

2.4.3 Graphics Objects

In ScicosLab the new graphics mode is the default graphics mode. The old graphics mode is still available, and in a Scilab session the user can simultaneously have graphical windows in both old and new modes.

The two modes share the high-level graphics functions defined in Table 2.6 but differ in their philosophy. The new graphics mode provides functions for accessing properties of graphics objects or changing the graphics object hierarchy. In the new graphics mode, it is not the graphics function calls that are recorded, but rather the state of the graphics is recorded.

The state of the graphics can be viewed as an instance of an object called a Figure. A Figure instance can be built by calling graphics functions. It contains in its fields the graphical hierarchy of objects that compose the displayed figure.

Thus a hierarchy of objects, which are called entities, compose a graphic Figure instance. Each entity has its own fields, which can also be entity instances. For example, the top-level object is always a Figure that contains a list of Axes (graphics subwindows). An Axes entity contains in its fields basic graphics objects like Polylines, Arc, ... A set of entities can be grouped using an Aggregation entity.

The main purpose of this new graphics mode is to give the user the possibility to access each entity in the object's hierarchy that composes a displayed graphic. By accessing we mean that the user can set or get object properties, for example the color of a displayed curve, using Scilab functions set or get. The displayed graphic is automatically updated to reflect the property changes.

As noted earlier, the properties of graphics objects can be changed through the Edit menu of graphics windows. This is particularly useful in taking a Scilab-generated figure and preparing it for use in presentations or publications. Properties such as color and style of lines and graphs are found under the properties of Polyline. A slide will adjust the color. The axis style and labeling are under the properties of Axes.

Thus, from the user point of view, a Figure can be viewed as an instance of a dynamic tree object. The nodes of the tree are called entities. The user can change the graph by adding or deleting nodes and changing the node properties. The Scilab name of a node pointer is a handle.

The two graphics modes can coexist in a Scilab session, but a default mode is set by the Scilab startup: set("old_style","off"|"on") can be used to set the default mode, and set("figure_style","new"|"old") can be used to open a graphical window in the specified mode.

We will now give some examples showing some of the graphics entities in action.

In this first example a graphics window is created, a default scale is set by calling plot2d, and a rectangle is drawn. In order to change the rectangle's properties, we first need to get a handle on it. Since the rectangle is the last drawn object, it is the current entity of the current graphics window, and in that case r=get("hdl") is the easiest way to access the rectangle. It is then possible to change the displayed rectangle's properties.

```
⟼ a=get("current_axes");            ← get a handle on the newly created axes
⟼ a.data_bounds=[-1,-1;1,1];        ← set the boundary values for x and z
⟼ xrect(0,0,0.5,0.5)                         ← draw a rectangle
⟼ r=get("hdl");              ← get the current node, i.e, the last created objects
⟼ r                                  ← r is a handle to a Rectangle entity
 ans =

Handle of type "Rectangle" with properties:
===========================================
parent: Axes
children: []
mark_mode = "off"
mark_style = 0
mark_size_unit = "tabulated"
mark_size = 0
mark_foreground = -1
mark_background = -2
line_mode = "on"
line_style = 0
thickness = 1
fill_mode = "off"
foreground = -1
data = [0,0,0.5,0.5]
```

```
visible = "on"
clip_state = "off"
clip_box = []
```

`⟼r.fill_mode="on";`	← changing the displayed rectangle's properties
`⟼r.foreground=5;`	
`⟼r_line_style=6;`	

In order to illustrate entities navigation, we can accomplish the same thing as the previous example in a more complex way by navigating, from node to node, in the graph of the objects associated with the figure.

`⟼h=get("current_figure");`	← handle to the default graphic window
`⟼ch = h.children;`	← the Axes entities
`⟼ch == get('current_axes')`	← the same ?
` ans  =`	

```
 T
```

`⟼r1= ch.children;`	← graphic entities
`⟼r1== r`	← we have found the rectangle again
` ans  =`	

```
 T
```

Since `get("current_figure")` and `get('current_axes')` are frequently used, they can be abbreviated as `gcf()` and `gca()`. One can also use `gdf()` and `gda()` to get a default figure and default axes, which are undrawn objects that are used as placeholders to store global default values used in creating respectively a new figure or new axes.

`⟼q=gda();`	← get the default axes
`⟼q.thickness=2;`	← default line thickness to 2
`⟼q.grid=[4,4];`	← I want a grid on my plot
`⟼q.font_size=4;`	← increase default font size
`⟼x=(0:0.1:2*%pi)';`	
`⟼plot2d(x,sin(x),rect=[0,-2,2*%pi,2]);`	
`⟼xdel();`	
`⟼sda();`	← reset the default axes to default values
`⟼plot2d(x,sin(x),rect=[0,-2,2*%pi,2]);`	← just check it

We have said that through properties one can change the graphical aspect of a displayed figure. But most of the time we want to set properties for undisplayed objects and only when all the properties are set to our requested values do we want to see the figure. This is easy to do since all the graphics entities have a `visible` property, which can be `on` or `off`. Moreover, we can start a graphics function call sequence by `drawlater` and end it with `drawnow` to control the visibility process. This is illustrated by the following script, which produces Figure 2.12.

```
u = %pi*(-1:0.2:1)/2;
v = %pi/2*(-1:0.2:1);
n = size(u,'*');
```

```
x= cos(u)'*exp(cos(v));                    ← a surface parameterized by u,v
y= cos(u)'*sin(v);
z= sin(u)'*ones(v);
col=ones(u)'*cos(v);                            ← a color for each vertex
col=(n-1)*(col-min(col))/(max(col)-min(col))+1;       ← rescale colors
drawlater();                                            ← draw later !
[xx,yy,zz]=nf3d(x,y,z);              ← from (x,y,z) to four sided faces
[xx,yy,zzcol]=nf3d(x,y,col);              ← change the colors as well
xx=[xx,-xx];yy=[yy,-yy];zz=[zz,zz];zzcol=[zzcol,zzcol];     ← symmetry
a=gca();                                  ← handle to the current axes
plot3d(xx,yy,list(zz,-zzcol));                         ← Let's do it
f=gcf();                                     ← handle to the figure
f3d=a.children;                ← just one child, it's our 3d plot
f3d.hiddencolor=n;                   ← set the hidden face's colors
a.rotation_angles = [55,110];               ← set the view angles
f.color_map= graycolormap(n);                  ← set the colormap
drawnow();                       ← show the result in a graphics window
```

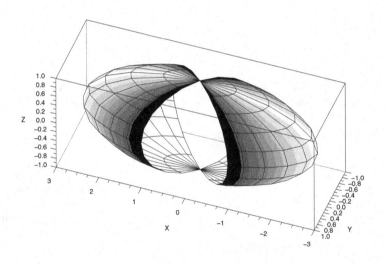

Figure 2.12. plot3d with the new graphic.

For those who prefer a graphical interface, keep in mind that an Edit button is available in the graphics window menu, which when launched opens up Tcl/Tk menus that can be used to navigate and change graphical object properties.

Tables 2.6, 2.7, and 2.8 summarize the Scilab graphics functions, graphics primitives, and graphics window primitives respectively.

2.4.4 Scilab Graphics and LaTeX

This section is for those interested in inserting Scilab graphics into LaTeX documents [27] (the typesetting system used to produce this book).

champ,champ1,fchamp	2D vector fields
contour,fcontour,contour2d,	
contourf, fcontour2d	2D Level curves
eval3d,eval3dp,genfac3d	3D utility functions for facets computation
fec	2D plot of a linear function on triangles
geom3d	from 3D coordinates to 2D projected ones
grayplot,fgrayplot	2D level curves as filled rectangles
hist3d, histplot	histograms
isoview	select isometric scales
legends	2D legends
Matplot,Matplot1	2D plot of matrices
param3d,param3d1	3D parametric curves
paramfplot2d	animation of 2D parametric curves
plot, plot2d[1-4],fplot2d	2D curves
plot3d,plot3d1,fplot3d,	
fplot3d1,plot3d2,plot3d3	surface drawing
polarplot	2D polar curves
Sfgrayplot, Sgrayplot	2D level curves (smoothed with fec)
subplot	split graphics window into subwindows

Table 2.6. Graphics functions.

xarc,xarcs	draw ellipses and circle
xarrows	draw a set of arrows
xaxis	axis drawing
xclea	clear a rectangular zone
xclip	fix a clip zone
xfarc,xfarcs	paint ellipses and circle
xfpoly,xfpolys	paint polylines
xfrect	paint a rectangle
xgrid	add a grid on 2D graphics
xnumb	draw numbers
xpoly,xpolys	draw a polyline
xrect	draw a rectangle
xrects	draw or paint a set of rectangles
xrpoly	draw a regular polygon
xsegs	draw a set of segments
xstring,xstringb	draw a string
xstringl	return the bounding box of a string
xtitle	add title to a graphic

Table 2.7. Graphics primitives.

Inserting a Scilab graphic in a LATEX document is an easy task since Scilab can export graphics as Postscript files and there are commands in LATEX to include Postscript files. Exporting a scilab graphic to Postscript can be, as already noted, achieved through the **File** menu of the graphics window, but it is also possible to use the Scilab command **xs2eps** (or **xs2ps** and the program **BEpsf**) to generate a Postscript file.

However, sometimes one wants to have a figure that is a mixture of Scilab graphics images and LATEX mathematical expressions. This can require translating the Postscript strings inserted in a Postscript document to LATEX commands in order, for example, to get mathematical expressions inserted in a graphic. There exists a LATEX package named

`driver`	select graphics driver
`graycolormap,`	predefined colormaps
`hotcolormap, ...`	
`xbasc`	clears graphics window and recorded graphics
`xbasimp`	sends Postscript graphics to printer
`xbasr`	redraws a graphics window
`xclear`	clears graphics window
`xclick,xgetmouse`	wait for mouse click
`xdel`	delete a graphics window
`xend`	stop a driver session
`xget`	get graphics context values
`xgetech`	get current scale
`xgraduate,graduate`	utility for graduation
`xinfo`	print a string in graphics window information widget
`xinit`	initialize a graphics driver
`xlfont`	add a font to the current driver
`xload`	saved graphics reloader (saved with `xsave`)
`xname`	give a name to current graphics window
`xpause`	pause
`xs2fig,xs2gif,`	
`xs2ppm,xs2ps`	save a graphics to file in `Xfig`, `gif`, `ppm`, or `Postscript` format
`xsave`	save graphics in a machine-independent format (use `xload` to reload)
`xselect`	raise a graphics window
`xset`	set graphics context values
`xsetech`	fix the current scale
`xtape`	recorded graphics management
`winsid`	return the graphics window's numbers

Table 2.8. Graphics window primitives.

`psfrag` that facilitates this task. Using `psfrag` you can substitute strings present in the Postscript file by LATEX expressions. The idea is to position strings in Scilab graphics using ASCII tags that can be substituted later by LATEX expressions. For example, in the following code the string `exprlatex` can be replaced by the LATEX expression

$$y = \frac{e^{-x^2/2}}{\sqrt{2\pi}}$$

using the LATEX code

```
\psfrag{exprlatex}{$y=\frac{e^{-x^2/2}}{\sqrt{2\pi}}$}
```

This is done by generating the figure in Scilab by

```
⊢→ v=rand(1,2000,'n');
⊢→ histplot([-6:0.4:6],v,[1],'015',' ',[-4,0,4,0.5],[2,2,2,1]);        ← histogram
⊢→ function [y]=f2(x); y=exp(-x.*x/2)/sqrt(2*%pi); endfunction;
⊢→ x=-6:0.1:6;x=x';plot2d(x,f2(x),1);                ← superposing an ℝ² curve
⊢→ xstring(2.3,0.2,'exprlatex');                        ← adding a string
```

Then after the Scilab figure is exported, in this case with the name `figpsfrag.eps`, the following LATEX commands generate Figure 2.13 in this book:

```
\begin{figure}
  \psfrag{exprlatex}{$y=\frac{e^{-x^2/2}}{\sqrt{2\pi}}$}
  \begin{center}
    \includegraphics[width=8cm]{code-book/figpsfrag}
  \end{center}
  \caption{Added \LaTeX\, expression with \tt{psfrag}.}
\end{figure}
```

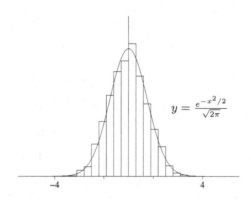

Figure 2.13. Added LaTeX expression with `psfrag`.

There is another solution for Scilab Postscript generated files since they contain, in the Postscript code, LaTeX information about the inserted strings (their positions and contents). It is easy to write a Scilab function to extract this information in order to use it at the LaTeX level. One such function is `xs2latex`, provided in Appendix B.3. Running this function, for example with the command `xs2latex(0,'figps.ps')`, produces two files `figps-ps.eps` and `figps-tex.tex`, which can be inserted in a LaTeX document as follows:

```
\begin{figure}[hbtp]
  \begin{center}
    \begin{picture}(0,0)
      \includegraphics[scale=1.0]{code-book/figps-ps.eps}
    \end{picture}
    \input{code-book/figps-tex.tex}
  \end{center}
  \caption{\label{figpslatex}Figure using \verb+xs2latex+.}
\end{figure}
```

Note that in order to change the default scale, it is not enough just to change `scale` in the code. The scale factor also has to be changed in `figps-tex.tex` by changing the `unitlength`.

Using the `xs2latex` macro, we can produce a figure with embedded LaTeX strings and mathematical expressions as shown in Figure 2.14.

The Scilab code used to produce Figure 2.14 is given below. This code also presents new graphics primitives in new graphics style:

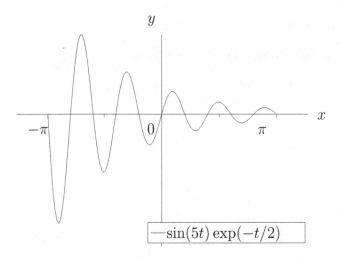

Figure 2.14. Figure obtained with xs2latex.

```
⊢→ t=-%pi:0.05:%pi;
⊢→ a=gca();
⊢→ a.font_size=5;
⊢→ plot2d(t,sin(5*t).*exp(-t/2))
⊢→ hl=legend(['$\sin(5 t)\exp(-t/2)$'],4);  ˙

⊢→ a.title.font_size=5;
⊢→ a.x_label.font_size = 5;
⊢→ a.y_label.font_size = 5;
⊢→ a.auto_ticks='off';
⊢→ a.box="off";
⊢→ a.x_location="middle";
⊢→ a.y_location="middle";
⊢→ xt=[-%pi,0,%pi];

⊢→ a.x_ticks= tlist(['ticks','locations','labels'],xt,...
                ["$-\pi$","$0$","$\pi$"]);
⊢→ xtitle("","$x$","$y$");

⊢→ xs2latex(xget("window"),"figps");
```

2.4.5 Old Graphics Style

As pointed out earlier, the old graphics style is considered dated and will not be maintained in the future. However, many examples and contributions to Scilab (the most important one being Scicos) still use the old graphics style, so we will give a brief presentation of the old graphics style here.

   In the old graphics style, the standard graphics output can be a graphics window, a Postscript file, or an Xfig file depending on the selected graphics driver. The current graphics driver is obtained with the `driver()` function. A driver is selected using the same function with the desired driver given as a string. For example, `driver('Pos')` selects Postscript. Alternatively, one can use the function `xinit` or use the Scilab graphic export menu that appears under the file button on the Scilab Graphics window.

When the selected driver is the one that redirects output to graphic windows, all Scilab graphics instructions are recorded sequentially. When a graphics window is resized or when a zoom is executed, Scilab replays the recorded instructions and produces a new graphic adapted to the size changes. In the same way, graphics instructions can be replayed after a driver change. For example, they can be used to obtain a Postscript version of a displayed graphics window. This can de done explicitly using the function `xs2ps`.

The default driver associated with a graphics window (the `Rec` driver) enables the recording of graphics instructions. Recorded instructions can be replayed using the function `xtape` or the function `xbasr`. Function `xbasc` can be used to clear the current window and to clear the associated recorded graphics instructions. Destroying a graphics window by clicking or by calling `xdel` also clears the associated recorded graphics.

It can sometimes be useful to use a graphics window without graphic recording, for example during animations. This is easily done by switching to the driver `X` (or `W`).

The default graphics behavior is that graphics display appears as soon as a graphics function is called. It is possible to attach a pixmap to a graphics window in order to redirect graphics outputs to that pixmap. The graphics window is redrawn using the pixmap status by a user specific command.

The pixmap mode of a graphics window is a window graphics parameter, and each graphics window has a large set of parameters. They define the graphics context of a graphics window. Many graphics windows can coexist, but only the graphics context of the current graphics window can be changed. The ID of the current graphics window (which is an integer) can be obtained using `xget('window')`, and changing the graphics window is done with `xset('window',n)`. The graphic context parameters of the current graphics window are obtained with the function `xget` and changed using the function `xset`.

The pixmap mode can be very useful, in particular for animation, so let us consider it more carefully. This mode is selected by calling `xset('pixmap',1)`. As explained earlier, when this mode is selected, graphics are directed to a pixmap. The content of the pixmap is copied to the graphics window when the `xset('wshow')` instruction is executed. The pixmap used for graphics is cleared by calling `wset('wwpc')`. This mode can be used for graphics animations since it avoids the flicker caused by having to clear a graphics window before redrawing it. The following script generates a movie of a rectangle that moves and changes its color pattern.

```
set old_style on
xsetech(frect=[0,0,10,10])
xrect(0,10,10,10)
n=100;
xset('pixmap',1)
driver('X11');
 for k=-n:n,
   a=ones(n,n);
   a= 3*tril(a,k)+ 2*a;
   a= a + a';
   k1= 3*(k+100)/200;
   Matplot1(a,[k1,2,k1+7,9])
   xset('wshow')
   xset('wwpc')
```

end
```
xset('pixmap',0)
```

The final point concerns scale computations. Graphics functions that aim at drawing simple objects, for example rectangles **xrects**, or polylines **xpoly**, use the current graphics scale. It is good scilab programming to set a graphics scale before calling such functions even if no error messages are provided by Scilab if the scales are not set. A current scale is set each time a high-level graphics function is called, such as **plot2d**, or when an explicit call to the function **xsetech** is used.

The default behavior of high-level functions is to adapt the current graphics scale so as to remain compatible with previous graphics function calls. But optional parameters can be used to change this default behavior. This is illustrated by the next example, which generates the graphs in Figure 2.15 and Figure 2.16.

```
⟼ xbasc();

⟼ t=linspace(0,2*%pi);
⟼ plot2d(t,sin(t))
⟼ plot2d(t,sinh(t/%pi))          ← the current scale changes so as to contain the two
                                    graphs (Figure~2.15)

⟼ [a,b,c,d]=xgetech();
⟼ b                               ← get the current scale [xmin,ymin,xmax,ymax]
  ans  =

!  0.  - 1.   7.    4. !

⟼ xbasc();

⟼ plot2d(t,sin(t))
⟼ plot2d(t,sinh(t/%pi),strf="000")  ← we don't want to change the current scale
                                       (Figure~2.16)

⟼ [a,b,c,d]=xgetech();
⟼ b                               ← get the current scale [xmin,ymin,xmax,ymax]
  ans  =

!  0.  - 1.   7.    1. !
```

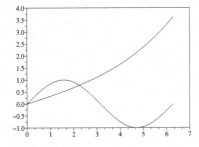

Figure 2.15. Automatic rescale.

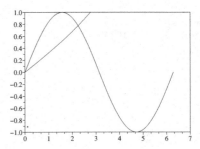

Figure 2.16. Using current scale.

2.5 Interfacing

In many areas, especially in modeling and simulation, there are computer programs written in other languages that we may want to interface with Scilab. Depending on the application area and the problem, these other programs could be written in C, C++, FORTRAN, etc. In principle, this interfacing could go either way. For example, we might want a FORTRAN program to use Scilab or for Scilab to use a FORTRAN program. We shall assume that the goal is for Scilab to be able to use another program. This section will discuss how this interfacing is done. There are two somewhat different scenarios in which such a program is to be used.

The first scenario is that the outside program is being called from within a Scilab function. For example, the outside program might return the value of the right-hand side of the differential equation that we wish to integrate with Scilab's ordinary differential solver `ode`. In that case, the interface is somewhat simpler and is described in Subsection 2.5.1.

In the second scenario we wish to have the given program become a Scilab function that we want to be able to call and use just like any other built-in Scilab function. This is discussed in Sections 2.5.2 and 2.5.3 it will often require some knowledge of the C programming language and the language used by the given external program.

2.5.1 Linking Code

Writing an interface is not always necessary in order to access external code. Suppose, for example, that you want to create a new Scicos block whose behavior is described by C-code. You do not need to access your block from the Scilab interpreter since the call will be performed directly by Scicos's internal simulator. In this case what you need is just to link your external code and update the Scilab internal table of dynamically linked functions. Then Scicos will be able to access the new code.

In fact, most of the Scilab functions that accept an `external` as argument can use the dynamic linking facility to access externally coded functions. We note `ode`, `dassl`, `optim`, and `fsolve`. Of course, when using this facility the calling sequence of the external function is constrained by the function it will be passed to. Each external-aware function has its own imposed calling sequence. There's also a special Scilab primitive called `call`, which can be used to call from the Scilab interpreter a dynamically linked routine that has in its arguments just integer, double, or character pointers. It can be used for testing external code. Just as in the previous section, we still have to create a shared library and load it in Scilab.

Linking external code can be done in two different ways. The function `addinter` is used to add new primitives to Scilab and requires the writing of interfaces. The function `link` just adds a new entry in the internal Scilab table of external functions. As pointed out earlier, such entries can be used by some Scilab primitives. Here we will use the `link` command through the `ilib_for_link` encapsulation. We encapsulate the call to `dlaswp` in another function to simplify the number of requested arguments:

```
#include <string.h>
#include "stack-c.h"

extern int C2F(dlaswp)(int *n,double *A,int *lda,int *k1,int *k2,int *ipiv,int *step);

int dlaswp1(double *A,int *m,int *n,int *mipiv,int *ipiv)
```

```
{
  const int un =1;
  C2F(dlaswp)(n,A,m,&un,mipiv,ipiv,&un);
}
```

We can compile the code using a small Scilab script called `builder.sce` (a more precise description of builders is given in Section 2.5.3) :

```
↦ names=['dlaswp1']; // entry points
↦
↦ files = 'dlaswp1.o'; // object files
↦ flag = 'c' ;
↦ ilib_for_link(names,files,[],"c");
   generate a loader file
   generate a Makefile: Makelib
   running the makefile
   compilation of dlaswp1
   building shared library (be patient)
```

In the previous code `flag='c'` is just there to specify that the function `dlaswp` is a C-coded function. This is used to guess the proper naming convention of entry points. We load the code in a running Scilab with the generated `loader.sce` file:

```
↦ // generated by builder.sce: Please do not edit this file
↦ // ------------------------------------------------------
↦ dlaswp1_path=get_absolute_file_path('loader.sce');
↦ link(dlaswp1_path+'libdlaswp1.so',['dlaswp1'],'c');
shared archive loaded
Link done
```

Now we can call the newly linked function not directly, but, for example, through the `call` function. The `call` function plays the role of the interface for a small subset of argument types. For example, we can call `dlaswp` as follows:

```
shared archive loaded
Link done

↦ perm=[1,3,4,3];mp=size(perm,'*');
↦ A=eye(4,4);[m,n]=size(A);
↦ y=call("dlaswp1",A,1,"d",m,2,"i",n,3,"i",mp,4,"i",perm,5,"i","out",1)
 y  =

!   1.    0.    0.    0. !
!   0.    0.    1.    0. !
!   0.    1.    0.    0. !
!   0.    0.    0.    1. !
```

The first call argument is the name of the function to be called. It is the name that was given to the link command as the entry point name. Other arguments from A to "out" give information on the arguments that are to be passed to dlaswp1. Then the arguments that follow the keyword "out" give information about returned values.

Each transmitted argument is described by three entries. For example, A,1,"d" means that the first argument of dlaswp1 is an array of double (precision) that points to the data of matrix A. The 1 that follows the "out" keyword means that the first returned value is also the first argument.

To provide a more explicit example, suppose that you want to find a zero of the nonlinear equation $\cos(x)x^2 - 1 = 0$. You can build a Scilab function:

⟼ **function** y=f(x)
⟼   y=cos(x)*x^2-1
⟼ **endfunction**

and using fsolve and the function f as its argument try to find a zero numerically:

⟼ y0=10; y=fsolve(y0,f)
 y   =

    11.003833

⟼ f(y)
 ans   =

    3.819D-14

However, we shall use this problem to illustrate linking. Suppose that the function f is C-coded (for example for efficiency) and it respects the calling convention of the Scilab fsolve primitive. One such implementation of f would be the following C-code:

```
#include <math.h>
void f(int *n,double *x,double *fval,int *iflag)
{
  *fval = cos(*x)*(*x)*(*x) − 1;
}
```

As before, we build and load a shared library using ilib_for_link. Then we pass the problem to solve to function fsolve by just giving a string that gives the name of the entry point to use, which in this problem is "f". The result is:

```
⟼ ilib_for_link('f','f.o',[],'c');
   generate a loader file
   generate a Makefile: Makelib
   running the makefile
   compilation of f
   building shared library (be patient)

shared archive loaded
Link done
```

⟼ y0=10; y=fsolve(y0,"f")
 y =

    11.003833

Many other Scilab functions can use C- and Fortran-coded functions directly, including ode, optim, etc.

2.5.2 Writing an Interface

The main purpose of this section is to show how to integrate C- compatible functions into Scilab. Suppose that we want to add new primitives to Scilab that will call external C-like functions. By C-like functions, we mean functions that can be coded in any language that remains compatible with C code in terms of linking and calling stack conventions. Examples include Fortran code and C++ ("extern C") functions.

We start with two examples that are Fortran-coded functions from the LAPACK library. [4] The first function is dgetrf, which performs an LU factorization of a given matrix. Given a m×n matrix A coded as an array of double-precision scalars of size m×n and an integer array of size Max(m,n), denoted by ipiv, the dgetrf routine performs an in-place LU factorization of a matrix. In fact, the LU factorization can be done with row interchanges. The integer array ipiv is here just to retain a coding of the row permutation. A Fortran subroutine can be called from C code with the convention that all arguments are transmitted through pointers. Note also that the naming conventions differ slightly from C to Fortran and are compiler-dependent. We enclose the Fortran name in a C2F(.) macro call, which will take care of naming conventions. Thus, the calling sequence for dgetrf from C-code is as follows:

extern int C2F(dgetrf)(**int** *m,**int** *n,**double** *A,**int** *lda,**int** *ipiv,**int** *info);

The second function we want to use is dlaswp. This routine simply applies a permutation described by an ipiv integer array to a given matrix:

extern int C2F(dlaswp)(**int** *n,**double** *A,**int** *lda,**int** *k1,**int** *k2,**int** *ipiv,**int** *step);

In order to access these two functions we add two Scilab primitives also called dgetrf and dlaswp with the following calling syntax:

```
[LU,perm]=dgetrf(A);
[C]=dlaswp(B,perm);
```

We still need to write glue code for converting Scilab arguments to C form, calling the external functions, and back-converting C objects to Scilab returned values. Most of the time the glue code is written in C and is called an interface, so that an interface is just a C-function. We need an interface for each function to be included into Scilab.

The two interfaces for dgetrf and dlaswp are as follows:

```
#include <string.h>
#include "stack-c.h"
```

extern int C2F(dgetrf)(**int** *m,**int** *n,**double** *A,**int** *lda,**int** *ipiv,**int** *info);

int int_dgetrf(**char** *fname)

```
{
  int l1, m1, n1, mipiv,nipiv=1,lipiv,info;
  CheckRhs(1,1);
  CheckLhs(1,2);
  GetRhsVar(1, "d", &m1, &n1, &l1);
  mipiv= Max(m1,n1);
  CreateVar(2, "i", &mipiv,&nipiv, &lipiv);
  C2F(dgetrf)(&m1,&n1,stk(l1),&m1,istk(lipiv),&info);
  if ( info < 0 )
    {
      Scierror(999,"%s: argument %d of dgetrf has an illegal value\r\n",
               fname,-info);
      return 0;
    }
  else if ( info > 0 )
    {
      sciprint("%s: U(%d,%d) is exactly zero\r\n",info+1,info+1);
    }
  LhsVar(1) = 1;
  LhsVar(2) = 2;
  return 2;
}


#include <string.h>
#include "stack-c.h"

extern int C2F(dlaswp)(int *n,double *A,int *lda,int *k1,int *k2,int *ipiv,int *step);

int int_dlaswp(char *fname)
{
  int l1, m1, n1, mipiv,nipiv,lipiv,info,un=1;
  CheckRhs(2,2);
  CheckLhs(1,2);
  GetRhsVar(1, "d", &m1, &n1, &l1);
  GetRhsVar(2, "i", &mipiv,&nipiv, &lipiv);
  mipiv= mipiv*nipiv;
  C2F(dlaswp)(&n1,stk(l1),&m1,&un,&mipiv,istk(lipiv),&un);
  LhsVar(1) = 1;
  return 1;
}
```

Note that in the two previous interfaces we have used functions, macros, or data from the interface library (CheckRhs, CheckLhs, GetRhsVar, CreateVar, and LhsVar). The header file stack-c.h is there to provide macro definitions and prototypes for the interface library functions.

At first sight, these two functions might seem complex, but most of the time we do not have to write the full interface from scratch. Picking pieces of code from the examples provided is the easiest way to write new functions. We won't describe in full detail all the functions that are present in the interface library. Rather we will try to give the rules that are to be followed for writing an interface.

How does the interface work? When at the Scilab level the user enters the command [LU,perm]=dgetrf(rand(5,5)), the arguments of the dgetrf function are first evaluated and their resulting values are stored in a stack (the function calling stack) in the order they appear in the calling sequence. Here we just have one argument; `rand(5,5)` is evaluated and the resulting matrix is stored on the calling stack at position 1. The Scilab calling sequence is also analyzed by the interpreter in order to evaluate how many returned arguments are expected. Here two returned arguments are expected.

The glue code for `dgetrf`, here called `int_dgetrf`, is then called by the interpreted code. The first job to be performed by the interface is to check that the number of given arguments is correct and that the number of expected returned arguments is also correct. This is performed by `CheckLhs` and `CheckRhs` . For example, if the number of given arguments is outside of the given bounds, then `CheckRhs` will print an error message and we return from the interface to the Scilab interpreter with a Scilab error raised.

In the interface each variable present on the stack is characterized by an integer that is its position on the stack. Calling arguments are, when entering the interface, stored on the stack from position 1 to n if n is the number of given arguments.

The next job to be performed is to check that each given argument has the expected type and proper dimensions. If the type is correct, then we need to give a pointer to the associated argument data. This pointer will be given to the C-function we are interfacing. The next command,

```
GetRhsVar(1, "d", &m1, &n1, &l1);
```

checks that the first given argument is a scalar matrix (`"d"`) and, in case of success, returns in `m1` and `n1` the matrix dimensions and in `l1` an identifier that can be used to get a pointer to the double array data (`stk(l1)`).

Before calling the function `dgetrf`, we need to allocate space in the Scilab calling stack for the integer array `ipiv`, which is to be of dimensions $1 \times \mathrm{Max}(m1,n1)$. The C-code

```
mipiv= Max(m1,n1);
CreateVar(2, "i", &mipiv,&nipiv, &lipiv);
```

will create a scalar matrix with the proper size. The `"i"` is here to say that the array is to be considered as an integer array. Data access is done through an integer pointer `istk(lipiv)`. Since the first available position is position 2, we use this position for our newly created variable.

The calling syntax of `CreateVar` is the same as the calling syntax of `GetRhsVar`, but note that in creating a variable, the dimension parameters are used as input data.

We are now ready to call the function `dgetrf`. After the call, the `Scierror` function can be used to raise a Scilab error and the `sciprint` function can be used to display warning messages or other relevant information.

After the call, we want to return the result of the LU factorization, which is stored in the double array `stk(l1)`, and the coded permutation, which is stored in `istk(lipiv)`. Notice that in `dgetrf` the double array `stk(l1)` is used as both input and output variable. However, there is no risk in modifying a Scilab input variable since when we are in the interface `GetRhsVar`, we do not have direct access to a given argument but only to a copy of that argument. It is thus safe to modify the `m1xn1` values of the double array `stk(l1)` and to use this variable as a Scilab return value.

We have checked at the beginning of the interface the number of requested returned values. At the end we just have to indicate at which location on the stack the returned arguments are to be found. The syntax `LhsVar(i)=j` is used to say that the `i`th value to be returned is at position `j` on the stack.

The second variable to be returned is the integer array. Note that by default Scilab matrices are stored as a double array and that we have used the data of the second variable

as an integer array. Since we have declared that the second variable was an integer array
("i"), Scilab will perform an automatic conversion of the array when returning from the
interface.

All interfaces are based on the same pattern, and a number of examples are available
in the subdirectories of SCI/examples. For example, the files included in the directory
SCI/examples/interface-tour-so give a tour of almost all of the functions that can be
used for interfacing. Note that it is possible, for example, to interface a function that itself
uses a function as a parameter and at the Scilab level we often want to pass a Scilab
function as an argument. It is also explained how to call the Scilab interpreter within an
interface to compute an intermediate value.

In the MATLAB world, interfaces are known under the name of mexfiles. As explained
in SCI/examples/mex-examples, mexfiles can be emulated in Scilab and can be used as
interfaces for Scilab functions.

2.5.3 Dynamic Loading

Writing glue code is just the first step. In order to use the desired code, we need to
compile the written code as a shared library. Usually this is .so files in Unix and .dll
files in Windows. Then we need to load it in Scilab, and tell Scilab the correspondence
between the interface names and the names of primitives to be added. Since this process
is system-dependent, we have developed Scilab script to hide system-dependent code from
the Scilab user. We have two interface functions that are coded in the two files intdgetrf.c
and intdlaswp.c. Note that there are no limitations or constraints on the number of files
and their names.

Using a text editor, we have to write a small Scilab script named builder.sce. This is
a canonic name and most Scilab users know that running builder.sce is used to set up a
contribution or an interface:

```
⊢→ // This is the builder.sce that must be run from this directory
⊢→ ilib_name  = 'liblapack';                        ← interface library name
⊢→ files = ['lapack.o'];                                 ← objects files
⊢→ libs  = [];                            ← other libraries needed for linking
⊢→ table =['dgetrf', 'int_dgetrf';     ← association table (scilab'name, interface-name)
⊢→       'dlaswp', 'int_dlaswp'];
⊢→ // do not modify below
⊢→ // -----------------------------------------------
⊢→ ilib_build(ilib_name,table,files,libs)
```

Of course, one can use as a model the builder.sce file contained in the directory
SCI/examples/interface-tutorial-so/builder.sce. It can be easily adapted to create a
new interface.

builder.sce provides the following information:

- ilib_name is a string that stands for the library name (a set of interfaces) that we are
  building.
- files is a string matrix that gives the object files that are needed to build the library. Note that the object files are suffixed with .o even under Windows. A unique
  builder.sce file is to be built and it should run as well under Unix and Windows.
- libs is a string matrix that gives a list of libraries (shared libraries) that are needed
  by the object files. Note that in our example, we are interfacing functions from the

LAPACK library, which is an already loaded Scilab executable. Thus we do not need to provide object files or libraries for providing the code of the `dgetrf` and `dlaswp` functions.

- `table` is a string matrix, where the association between a Scilab name and an interface name is provided. For example, its first line `'dgetrf', 'int_dgetrf';` indicates that the Scilab function `dgetrf` is implemented via the interface `int_dgetrf`.

Running the `builder.sce` script from Scilab will create a shared library and a set of new files.

```
// This is the builder.sce that must be run from this directory
ilib_name = 'liblapack';                              ← interface library name
files = ['lapack.o'];                                 ← objects files
libs  = [];                                  ← other libraries needed for linking

table =['dgetrf', 'int_dgetrf';     ← association table (scilab name, interface-name)
        'dlaswp', 'int_dlaswp'];

// do not modify below
// ----------------------------------------------
ilib_build(ilib_name,table,files,libs)
generate a gateway file
generate a loader file
generate a Makefile: Makelib
running the makefile
compilation of lapack
building shared library (be patient)
```

Among the created files, the most important one is a new Scilab script called `loader.sce`, which can be used when we need to load the shared library in a running Scilab. The `loader.sce` uses the Scilab function `addinter` to load the shared library and update the primitive table.

In practice, one executes the `builder.sce` once. Note that when running a `builder.sce` script the current Scilab directory must be the directory containing the `builder.sce` script, and the library is loaded with the `loader.sce` script each time we start a new Scilab session.

```
// generated by builder.sce: Please do not edit this file
// -------------------------------------------------------
liblapack_path=get_file_path('loader.sce');
functions=[ 'dgetrf';
            'dlaswp';
];
addinter(liblapack_path+'/liblapack.so','liblapack',functions);
shared archive loaded
```

Among the other generated files we note:

- A file named after the library name `liblapack.c`, called a "gateway" file, is also generated. It contains the main entry point of the library, which is used by the `addinter` function.

- A makefile named `Makelib`, which is system-dependent and, which can be used to rebuild the shared library. This makefile can be very useful for debugging purposes when the shared library build process fails.

We can now test the added primitives:

```
⟼ exec ('loader.sce');
```

```
shared archive loaded
```

```
⟼ A=rand(5,5);
⟼ B=A;
⟼ [LU,perm]=dgetrf(A);                                        ← call dgetrf
⟼ [E]=dlaswp(eye(A),perm);           ← apply permutation to identity matrix
⟼ L=tril(LU,-1)+eye(LU);U=triu(LU);                  ← separate L and U
⟼ if norm(L*U-E*A) > 10*%eps  then pause; end ;           ← check result
```

3

Modeling and Simulation in Scilab

One of the fundamental problems in many areas of science and engineering is the problem of modeling and simulation. Scilab provides a large array of tools for developing and simulating models of several types. For several of these tools it is possible to use them with abbreviated commands and default values of some parameters. However, to know how to choose the appropriate tools and how to get the kind of answers desired, it is often necessary to know something about how the algorithms are set up and what to do if there are difficulties.

In this chapter we shall describe the types of models that are available in Scilab and some of the tools available for their simulation. We shall also give some general comments about how to use the tools and how to choose between the tools.

3.1 Types of Models

3.1.1 Ordinary Differential Equations

One of the most popular types of model is that of ordinary differential equations (ODE) such as

$$\dot{y} = f(t, y). \tag{3.1}$$

Here y and f may be vector valued, y is a function of the real variable t, and $\dot{y}$ denotes the derivative of y with respect to t. System (3.1) is a first-order equation since it involves only first-order derivatives.

Some models are initially formulated in terms of higher derivatives, but they can always be rewritten as first-order systems by the introduction of additional variables. The Scilab tools assume that the differential equation has first been rewritten to be of first order. The exception to this is boundary value problems, where the numerical methods allow for higher-order differential equations.

Example 1. The system of second-oder differential equations

$$\ddot{y}_1 = \dot{y}_1 - y_2 + \sin(t), \tag{3.2a}$$

$$\ddot{y}_2 = 3\dot{y}_1 + y_1 - 4y_2, \tag{3.2b}$$

can be rewritten as

S.L. Campbell et al., *Modeling and Simulation in Scilab/Scicos with ScicosLab 4.4*,
DOI 10.1007/978-1-4419-5527-2_3, © Springer Science+Business Media, LLC 2010

$$\dot{y}_1 = y_3, \tag{3.3a}$$

$$\dot{y}_2 = y_4, \tag{3.3b}$$

$$\dot{y}_3 = y_3 - y_2 + \sin(t), \tag{3.3c}$$

$$\dot{y}_4 = 3y_3 + y_1 - 4y_2. \tag{3.3d}$$

The differential equation (3.1) generally has a family of solutions. To perform a simulation, additional information must be provided to make the solution unique. An initial condition is the value of y at a particular value of t. A boundary condition is when more than one value of t is involved in the additional conditions. An example of a boundary condition is $y(0) - 2y(\pi) = 0$. We will discuss boundary value problems later.

The basic existence and uniqueness result for differential equations in the form (3.1) says that if we specify $y(t_0)$, and f and $f_y = \partial f / \partial y$ are continuous functions of t and y near $(t_0, y(t_0))$, then there is a unique solution to (3.1) satisfying the initial condition. The solution will continue to exist as long as these assumptions hold. Another important fact is that the more derivatives of $f(t, y)$ that exist and are continuous, the more derivatives y has.

This may seem to be an esoteric fact, but it actually plays an important role in simulation. The reason is that most simulation tools try to deliver a requested accuracy and the error estimates are based on assumptions about the number of derivatives of the solutions. Suppose now that we have a controlled differential equation

$$\dot{y} = f(t, y, u(t)) \tag{3.4}$$

with $u(t)$ a given function. Then even if f has all the derivatives that one wants, y can have only one more derivative than u. Thus if a control is only piecewise continuous, so that it has jumps, which often occurs in applications, the solution y will not have a continuous first derivative at the jumps. This in turn can lead to serious problems with many numerical tools unless the behavior of u is taken into account when one is requesting the simulation.

3.1.2 Boundary Value Problems

Boundary value problems (BVP) are differential equations but with the information given at two or more times rather than at just one time. Two-point boundary value problems take the general form of

$$\dot{y} = f(t, y), \quad t_0 \le t \le t_f, \tag{3.5a}$$

$$0 = B(y(t_0), y(t_f)). \tag{3.5b}$$

If y in (3.5a) is n-dimensional, then there usually need to be n equations in (3.5b) in order to uniquely determine a solution. However, the theory is more complicated than that for initial value problems, as illustrated by the next example.

Example 2. Suppose that we have the boundary value problem

$$\dot{y} = \begin{pmatrix} 0 & 1 \\ -1 & 0 \end{pmatrix} y + \begin{pmatrix} 0 \\ 1 \end{pmatrix}, \tag{3.6a}$$

$$0 = \begin{pmatrix} y_1(0) \\ y_2(t_f) \end{pmatrix}. \tag{3.6b}$$

The solution of this differential equation is

$$y_1 = c_1 \cos(t) + c_2 \sin(t) + 1, \tag{3.7a}$$

$$y_2 = -c_1 \sin(t) + c_2 \cos(t), \tag{3.7b}$$

where c_1, c_2 are arbitrary constants. The first boundary condition (3.7a) gives $c_1 = -1$. The second boundary condition (3.7b) then gives that

$$0 = \sin(t_f) + c_2 \cos(t_f). \tag{3.8}$$

But (3.8) has a solution for c_2 only if $\cos(t_f) \neq 0$. Thus we see that the existence of a solution to this boundary value problem depends on the particular value of t_f. Here t_f cannot be $\frac{\pi}{2} + m\pi$, where m is an integer.

One of the nice features about ODEs is that the given initial conditions are at one point. This means that with the numerical methods given below the computation can be local and move from one step to the next. This makes it easier to solve problems on longer time intervals or with higher state dimension. With BVP, since the information occurs at more than one time, it is often necessary to use a more global algorithm that takes into account the full t interval. This leads to much larger systems of equations that have to be solved.

3.1.3 Difference Equations

The second major class of models is that of difference equations. These problems occur when there a quantity whose values are of interest only at discrete-time values or the values change only at discrete times. Difference equations can occur either because the process is naturally discrete and only undergoes changes at isolated times or because it is continuous but our observations occur only at isolated times. Also, many numerical methods for differential equations actually solve an approximating difference equation.

In difference equations there is an integer variable, which we denote by k. The solution is a sequence $y(k)$ and it has to solve a difference equation

$$y(k + 1) = f(k, y(k)), \quad y(k_0) = y_0. \tag{3.9}$$

Sometimes it is more natural to think of a sequence of times t_k, $k \geq k_0$, and a difference equation

$$z(t_{k+1}) = g(t_k, z(t_k)), \quad z(t_{k_0}) = z_0. \tag{3.10}$$

If the t_k are evenly spaced, so that $t_{+1} - t_k = h$ where h is a constant, then we can rewrite (3.10) in the form of (3.9) as follows:

$$v(k + 1) = g(w(k), v(k)), \quad v(k_0) = z_0, \tag{3.11a}$$

$$w(k + 1) = w(k) + h, \quad w(k_0) = t_{k_0}. \tag{3.11b}$$

Note that $v(k)$ of (3.11a) is the same as $z(t_k)$ from (3.10).

For the remainder of this section we will assume that the difference equation is in the form (3.9). The theory for solutions of difference equations is simpler than that for differential equations. Given a starting time k_0 and a value of y_0 that gives the initial condition $y(k_0) = y_0$, the values of $y(k)$ for $k > k_0$ are computed recursively using (3.9) and will exist as long as $(k, y(k))$ is in the domain of f.

There is one important difference between the discrete and continuous theories that is worth mentioning. The uniqueness theorem for differential equations tells us that if two solutions start at the same time but with different initial values and if the continuity assumptions on f, f_y are holding, then the solutions can never intersect. This is not true for difference equations.

Example 3. The difference equation (3.12) is at the origin after two time steps no matter where it starts at k_0:

$$y(k+1) = \begin{pmatrix} 1 & -1 \\ 1 & -1 \end{pmatrix} y(k). \tag{3.12}$$

The difference equations (3.9) and (3.12) are of first order since there is maximal difference of one in the values of k that are present. Just as systems of differential equations can always be written as first-order differential equations, difference equations can always be written as first-order difference equations.

Example 4. The equation (3.13) is of third order since $(k+3) - k = 3$:

$$y(k+3) = 4y(k+2) - y(k+1)y(k). \tag{3.13}$$

The difference equation (3.13) can be rewritten as a first-order difference equation by introducing additional variables as shown in (3.14):

$$y_1(k+1) = y_2(k), \tag{3.14a}$$
$$y_2(k+1) = y_3(k), \tag{3.14b}$$
$$y_3(k+1) = 4y_3(k) - y_2(k)y_1(k). \tag{3.14c}$$

The Scilab tools for working with difference equations assume that they have been written in first-order form.

3.1.4 Differential Algebraic Equations

Many physical systems are most naturally initially modeled by systems composed of both differential and algebraic equations [12]. These systems take the general form of

$$F(t, y, \dot{y}) = 0 \tag{3.15}$$

and are often called DAEs.

Example 5. One example of a DAE is a model of a constrained mechanical system where the constraint is on the position variables. Such a system takes the general form

$$\dot{x} = v, \tag{3.16a}$$
$$\dot{v} = G(x, v, t) + Q(x, t)\lambda, \tag{3.16b}$$
$$0 = H(x, t). \tag{3.16c}$$

System (3.16) is a DAE in $y = (x, v, \lambda)$. Equation (3.16c) gives the constraints. Also $Q(x, t) = H_x(x, t)^T$, and $Q(x, t)\lambda$ gives the force generated by the constraint.

Given a DAE model, there are two options. One option is to try to rewrite the DAE into an ODE or a simpler DAE. This is usually done by a mixture of differentiation and algebraic manipulations. The other option is to try to simulate the DAE directly.

The theory for DAEs is much more complex than the theory for ODEs. One problem is that there will exist solutions only for certain initial conditions that are called consistent initial conditions. Considerably more detailed information about DAEs is in [12]. Here it is enough to know two things. First, the structure of the DAE is important, and second, there is a nonnegative integer called the index. An ODE is is said to have index zero.

Example 6. The simple system

$$\dot{y}_1 = y_1 - \cos(y_2) + t, \tag{3.17a}$$

$$0 = y_1^3 + y_2 + e^t, \tag{3.17b}$$

is a DAE. From (3.17b) we see that the initial conditions must satisfy $y_1(t_0)^3 + y_2(t_0) + e^{t_0} = 0$, so that only some initial conditions are consistent. This particular example can be reduced to an ODE by solving (3.17b) for y_2 and substituting into (3.17a) to get an ODE in y_1.

Index-one DAE systems occur frequently and are the ones for which Scilab provides tools. They are also the ones that Scicos is able to solve when they occur in the form of implicit blocks. Index-one systems in the form (3.15) have the additional property that $\{F_{\dot{y}}, F_y\}$ is an index-one matrix pencil all along the solution and $F_{\dot{y}}$ has constant rank. Two important types of index-one DAEs are the implicit semiexplicit and semiexplicit index-one systems. The implicit semiexplicit index-one systems are in the form

$$F_1(t, y_1, y_2, \dot{y}_1) = 0, \tag{3.18a}$$

$$F_2(t, y_1, y_2) = 0. \tag{3.18b}$$

Here $\partial F_1 / \partial \dot{y}_1$ and $\partial F_2 / \partial y_2$ are both nonsingular. We shall sometimes refer to y_1 as a differential variable and y_2 as an algebraic variable.

The semiexplicit index-one DAE takes the form

$$\dot{y}_1 = F_1(t, y_1, y_2) \tag{3.19a}$$

$$0 = F_2(t, y_1, y_2) \tag{3.19b}$$

with $\partial F_2 / \partial y_2$ nonsingular. The DAE (3.17) is semiexplicit of index one. Index-one DAEs that are not semiexplicit or linear in the derivative terms are called fully implicit.

If (3.15) is to be solved by a numerical method such as `dassl` which is described in Section 3.2.4, then the numerical integrator needs an initial value of both y and $\dot{y}$. There are two difficulties. As noted, only some values of y are consistent. In addition, since $\partial F / \partial \dot{y}$ is singular, (3.15) only partially determines $\dot{y}$.

To illustrate, consider the problem of initializing (3.17) at $t = 0$. $y_1(0)$ and $y_2(0)$ must satisfy (3.17b). One consistent choice is $y_1(0) = -1$, $y_2(0) = 0$. Then $\dot{y}_1(0) = -2$ from (3.17a). But there is no obvious choice for $\dot{y}_2(0)$. One can either try to compute $\dot{y}_2(0)$ from additional information about the problem or utilize some of the utilities of the integrator. The second option may be the only one for a complex problem and will be illustrated when we discuss `dassl`. For the simple problem (3.17) we can differentiate (3.17b) and evaluate at $t = 0$ to get an equation that yields $\dot{y}_2(0)$.

3.1.5 Hybrid Systems

Many systems involve a mixture of discrete- and continuous-time events. These types of systems are generally called hybrid systems. The discrete nature of the system can occur in several different ways. These differences have to be taken into account in the simulation. The times at which a discrete variable changes value are sometimes called events.

When an event occurs, there may be a change in the differential equations. The state dimension may also change. It then becomes important to determine new initial conditions. This is referred to as the initialization problem. Sometimes the event occurs when some quantity reaches a certain value. For example, in the simulation of a mechanical system, if contact is made with an obstacle, then the equations change. It thus becomes necessary for the simulation software to be able to determine when the event occurs. This is known as root-finding ability.

Even when an event does not change the dimension of the state or the form of the equations, it can interfere with the error control of the integrator being used unless care is taken.

The Scicos software described in Part 2 of this book was specifically designed to handle hybrid systems. The new version released with Scilab 3.0 is able to simulate a variety of DAE hybrid systems.

3.2 Simulation Tools

The simulation tools in Scilab generally take three forms. One of these comprises the primary tools used by knowledgeable users on challenging problems. The second is simplified versions of the primary tools that are easier to use and are designed for simpler problems. The third is tools developed for special cases that occur often in certain areas of science and engineering. We will usually present them in the order of simplified, primary, specialized.

3.2.1 Ordinary Differential Equations

The primary simulation tool for ordinary differential equations is `ode`. In order to simulate a differential equation (3.1), the least that we can give the integrator is the initial time t_0, the initial value y_0, and a formula for evaluating $f(t, y)$. When we do this, the software picks the method and the error tolerances and adjusts the step size to try to meet the error tolerances. This will often result in many more values of the solution than are needed. Also, sometimes the value of the solution is desired at time values that are not time values used in the numerical simulation. Thus the final thing the simplified version needs is a vector of times at which the value of the solution is desired. The final entry of this vector is the final time for the integration. Thus the simplified calling sequence for `ode` is

$\longmapsto$ y=ode(y0,t0,t,f)

Here `f` is an external variable or string providing the function f that is the right-hand side (RHS) of (3.1). If $t = [t_0, t_1, \ldots, t_n]$, then y= $[y(t_0), y(t_1), \ldots, y(t_n)]$, where $y(t_i)$ is the estimate for the solution at time t_i. If `t` is a single number, then it is the final time of the integration. In earlier versions of Scilab, the function `f` could be defined either as a `.sci` file or on-line using the `deff` function. With version 3.0 of Scilab, functions may be defined on-line using the same syntax as if they were defined in a file. Thus if we wish to define the function $f(t, y) = -y + \sin(t, y)$ on-line, we can just write

$\longmapsto$ **function ydot = f(t,y)**
$\longmapsto$ **ydot=-y + sin(t)**
$\longmapsto$ **endfunction**

Unfortunately, the simple call of `ode` does not always work. More direct control of the simulation comes through two methods. One is the expanded or full version of the `ode` command. The other uses `odeoptions`.

In the full version of `ode` one can choose the method, specify the integration tolerances, provide the Jacobian if it is needed by the type of method chosen, and set up some other variables required by some methods. `ode` provides a standard interface to a number of solvers, in particular to those in ODEPACK[14, 32, 53]. CVODE from the SUNDIALS suite [21, 33] can be accessed by using Scicos since CVODE is the ordinary differential equation solver used in ScicosLab Scicos.

In choosing the method, several factors need to be taken into account. For problems with frequent restarts, there are advantages to one step methods such as the Runge–Kutta formulations over the multistep methods such as the BDF methods and Adams methods. The order of a numerical method is an indication of the error relative to the step size. Thus a fourth-order method would be expected to have error proportional to h^4 (assuming that a constant step h is used, and the steps are sufficiently small). On the other hand, multistep methods, for a given order, often require less computation. Some problems are referred to as stiff. There are lots of definitions of a stiff differential equation, but we will take stiff to mean that if one had an explicit method, then the step sizes would have to be reduced to ensure stability of the error propagation rather than to get the desired accuracy. Stiffness is often caused by large negative eigenvalues. Generally, implicit methods can take bigger time steps than explicit methods on stiff problems. Higher order is usually desirable. However, for implicit methods the higher the order, the smaller the stability region. This means that the higher-order method might have to take smaller steps than expected. This is especially true for problems that have highly oscillatory solutions. In addition, it does not make sense to go to high order if the equations, or input functions, do not have enough continuous derivatives.

The previous comments only touch on some of the issues involved in the choice of numerical method. Our experience has been that when the simple call does not work, then it is often better to alter the `atol` and `rtol` and use `odeoptions`, which is described shortly, rather than altering the method. The exception is with DAEs as discussed later.

The full calling sequence of `ode` is either of

```
[y,w,iw]=ode([type],y0,t0,t [,rtol [,atol]],f [,jac] [,w,iw])
[y,rd,w,iw]=ode("root",y0,t0,t [,rtol [,atol]],f [,jac],ng,g [,w,iw])
```

Any variables inside ode with square brackets are optional. Here `type` is a character string and refers to the type of method chosen. The available types are as follows:

`lsoda`: This is the default integrator. It is from the package ODEPACK. It automatically selects between a nonstiff predictor-corrector Adams method and the stiff Backward Differentiation Formula (BDF) method. Both methods are multistep methods. `lsoda` uses the nonstiff method initially and dynamically monitors data in order to decide which method to use. It also automatically varies the order.

`adams`: This calls the lsode solver from ODEPACK and uses the Adams method. It is for nonstiff problems.

`stiff`: This is for stiff problems. It calls the lsode solver from ODEPACK and uses the BDF method.

`rk`: This is an adaptive Runge-Kutta of order 4 (RK4). Thus it is a one-step method and can usually use larger steps after a restart than a multistep method can use.

`rkf`: This is another Runge–Kutta method by Shampine and Watts and is based on Fehlberg's Runge–Kutta pair of order 4 and 5 (RKF45). The method is of order 4. It simultaneously carries out a fifth-order method that is used for error prediction. This method is for nonstiff and mildly stiff problems when derivative evaluations are inexpensive. This method should generally not be used when the user demands high accuracy.

fix: This is the same solver as rkf, but the user interface is very simple. Only the rtol and atol parameters can be passed to the solver. This is the simplest method to try.

root: This is the lsodar solver with root-finding capabilities from ODEPACK. It is a variant of the lsoda solver. root finds the roots of a given vector function. We will discuss this more carefully below when we discuss ode_root.

discrete: We will discuss this option when we discuss discrete simulation tools below. See also ode_discrete for more details.

Two parameters that are especially useful are rtol and atol, which are real constants or real vectors of the same size as y. Their entries are the absolute and relative error tolerances requested for each entry of y. Setting these tolerances too tight can increase computational time and, in some cases, lead to integrator failure when tolerances cannot be met. Setting the tolerances too loose can lead to inaccurate answers. It should be kept in mind that the step size decisions are based on estimates of the error. The actual error in the solution could be higher. Default values for rtol and atol are rtol=1.d-5 and atol=1.d-7 for most solvers and rtol=1.d-3 and atol=1.d-4 for rfk and fix.

The parameter jac is an an external (function or character string or list) that provides a Jacobian for those methods that can utilize analytic Jacobians. These include the BDF based and other implicit methods for stiff problems. If jac is a function, then the syntax should be J=jac(t,y). J should evaluate to the matrix that is $\partial f / \partial x$.

Optional arguments w and iw are vectors for storing information returned by the integration routine. When these vectors are provided in the right-hand side of ode the integration restarts with the same parameters as in its previous stop.

When ode is called, the first thing it does is to check whether the variable %ODEOPTIONS exists. If %ODEOPTIONS does not exist, ode uses the default values of a number of parameters. %ODEOPTIONS can be set directly. Alternatively, these parameters can be changed using the function odeoptions:

⟼   %ODEOPTIONS=odeoptions()

A window then opens in which any desired parameter values can be entered. Parameters will stay at these values throughout the rest of the session unless they are reset or cleared.

The variable %ODEOPTIONS has the following entries:

```
%ODEOPTIONS=[itask,tcrit,h0,hmax,hmin,jactyp,mxstep,maxordn,maxords,...
            ixpr,ml,mu]
```

with default value [1,0,0,%inf,0,2,500,12,5,0,-1,-1]. The parameters in %ODEOPTIONS have the following uses and meaning.

itask: Takes on integer values in $[1,5]$. 1 is normal computation at specified times. 2 is computation at mesh points that are given in the first row of output of ode. 3 is one step at one internal mesh point and return. 4 is normal computation without overshooting tcrit. 5 is one step, without passing tcrit and return. Options 4 and 5 are used when there is critical time past which the differential equation changes so that one wants to integrate up to tcrit but not use information past that time.

tcrit: Assumes that itask equals 4 or 5 as described above. tcrit is a scalar.

h0: This is the first step tried by the integrator. It might be used with a Runge–Kutta method where you know that the solution starts out changing slowly and using a small first step wastes CPU time.

hmax: This is the largest step size that the integrator is allowed to use. One use of this is with simulations where there are long slow stretches and then rapid changes in the solution. Sometimes the integrators will start taking such large steps that they are unable to recover from a rapid change, or even worse, they miss an isolated event completely. This is illustrated in Example 7. If `hmax` is set small, it may be necessary to increase `mxstep`.

`hmin`: This is the minimum step size that can be accepted. Setting it greater than zero can make the error estimates less reliable and may result in a less-accurate solution. However, it also has the effect of forcing the integrator to keep going rather that letting the integrator take very tiny steps. This is most useful when there are local conditioning problems or loss of smoothness that is interfering with the error estimates.

`jactype`: This specifies how any nonlinear equations in the integrator will be solved. 0 is functional iterations with no Jacobian used. This is for `'adams'` or `"stiff"` only. 1 is a user-supplied full Jacobian. 2 is an internally generated full Jacobian. This is done by differencing and so is an approximate Jacobian. 3 is an internally generated diagonal Jacobian. This is for `'adams'` or `"stiff"` only. 4 is a user-supplied banded Jacobian (see `ml` and `mu` below). 5 is an internally generated banded Jacobian (see `ml` and `mu` below). Banded Jacobians naturally occur with some problems, such as the method of lines solutions of PDEs. The solvers can exploit the banded structure if they are informed about it.

`mxstep`: This is the maximum number of time steps the integrator is allowed to take. The default value is 500. Having a finite value here is important since it avoids having the algorithm spending a very long time on integration without permission. This can happen on longer intervals that require very tiny timesteps, or in integrating until a root is found but the surface is never crossed. 500 is probably a bit low for a default value and is there for historical reasons.

`maxordn`: This is the maximum nonstiff order allowed. It can be at most 12. The higher orders have some advantages, but they also have a number of disadvantages on some problems, so one should not always assume that higher is better. Setting `maxordn` to be less than 12 is another way to encourage the method to use smaller step sizes.

`maxords`: This is the maximum stiff order allowed. It is at most 5 since BDF methods are not stable at orders higher than 5. Setting `maxords=1` gives an implicit Euler's method. If one is having trouble integrating highly oscillatory systems, then it is often helpful to set `maxords` to 2 or maybe 3.

`ixpr`: This is the print level and is 0 or 1.

`ml, mu`: If `jactype` equals 4 or 5, then `ml` and `mu` are the lower and upper half-bandwidths of the banded Jacobian. The band is the `i,j`'s with $i-ml \leq j \leq ny-1$. If `jactype` equals 4, then the Jacobian function must return a matrix J that is `ml+mu+1*ny`, where `ny` is the dimension of $\dot{y}$ in $\dot{y} = f(t, y)$. Column 1 of J is made of `mu` zeros followed by $\partial f/\partial y_1$ (`1+ml` possibly nonzero entries). Column 2 is made of `mu-1` zeros followed by $\partial f/\partial y_2$, etc.

Example 7. As an illustration of the need for `odeoptions()`, we consider the problem of simulating

$$\dot{y} = -0.1y + g(t), \ y(0) = 1, \ 0 \leq t \leq 600, \tag{3.20}$$

with $g(t)$ zero except between 488.3 and 488.9, where g takes the value of 2. This differential equation is representative of the situation in which there are sudden events after a long period during which there is little change in the solution of the differential equation. One way to write g mathematically is

$$g(t) = 0.5(1 + \text{sign}(t - 488.3))(1 - \text{sign}(t - 488.9)).$$

The following Scilab script runs a simulation of (3.20) with the default settings of %ODEOPTONS. The result is plotted in graphics window 0. It then changes hmax to 0.1 and increases mxstep to 10,000 and plots the new solution in graphics window 1.

```
function z=g(t)
  z=0.5*(1+sign(t-488.3))*(1-sign(t-488.9))
endfunction

function ydot = f(t,y)
  ydot=-0.1*y+g(t)
endfunction

tt=0:0.1:600;
%ODEOPTIONS=[1,0,0,%inf,0,2,500,12,5,0,-1,-1];
y=ode(1,0,tt,f);
xbasc()
plot2d(tt,y);                                          ← Fig. 3.1

%ODEOPTIONS=[1,0,0,0.1,0,2,10000,12,5,0,-1,-1];
xbasc()
y=ode(1,0,tt,f);
plot2d(tt,y);                                          ← Fig. 3.2
```

If this script is called longstep.sci and is in the directory /Users/scampbell, then the simulation is carried out by

⟼ exec("/Users/scampbell/longstep.sci")

The simulation using ode with the no limit on hmax gives the graph in Figure 3.1. The effect of g is totally missing. This is because ODE integrators often take large steps during "quiet periods" and then sometimes totally miss events happening over relatively short periods. Changing hmax from %inf to 0.1 and mxstep to 10,000, we get the graph in Figure 3.2. The change in mxstep was needed because the new value of hmax would require more than the default value of 500 steps.

A careful discussion of all the factors in choosing between methods is a book in itself. The interested reader is referred to the classic volumes [30, 31]. But one question is, suppose that I am going to use default values. Should I worry about which method I use? The following example addresses the effect of method on time of computation. Actual timings are somewhat machine-dependent. We do not consider the important question of accuracy here but note that in the computations below, when we had the solutions graphed they appeared the same.

Example 8. Listed below is a Scilab script compode.sci, which defines several functions. The function comparison(a,b,c,d) generates an 8-dimensional linear system $\dot{y} = Ay + g$. The eigenvalues of A are $a \pm bi, a \pm bi, c, c, d, d$, and all modes appear in all entries of y. This differential equation is then integrated using the default method and adams, stiff, rkf, and the DAE integrator dassl discussed later in this chapter. The output is the computational times for each method in seconds. The times are found using tic and toc.

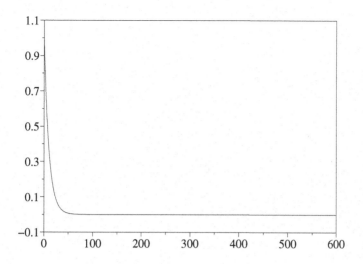

Figure 3.1. Simulation using default values of `hmax` and `mxstep`.

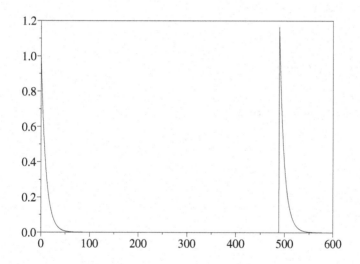

Figure 3.2. Simulation after using `odeoptions()` to change f, `hmax`, and `mxstep`.

```
function w=g(t)
  w=[zeros(1,6),cos(t),3*sin(2*t)]';
endfunction
function ydot=f(t,y)
  ydot=A2*y+g(t);
endfunction
function [r,ires]=res(t,y,ydot)
  r=ydot-A2*y-g(t)
  ires=0
endfunction

function zz=comparison(a,b,c,d)
  A=[a,-b,0,0;b, a, 0,0;0,0,c,0;0,0,0,d];
  II4=eye(A);Z=zeros(A);
  AA=[A,zeros(A);zeros(A),A];
  Q=[1 3 2 0 -1 4 6 0 10 4 5 7 8 -3 6 -5 ];
  Q=[Q,1,3,Q,3,6,8,9,Q,11,7,2,4,0,-3,6,3,9,-3];
  QQ=matrix(Q,8,8);
  A2=inv(QQ)*AA*QQ;
  tt=[0:0.1:100];t0=0;
  y0=[1 1 -1 1 2 0 -1 -1 ]';
  ydot0=A2*y0+g(0);
  tic;yy1=ode(y0,t0,tt,f);
  t1=toc();
  tic;yy2=ode("adams",y0,t0,tt,f);
  t2=toc();
  tic;yy3=ode("stiff",y0,t0,tt,f);
  t3=toc();
  tic;yy4=ode("rkf",y0,t0,tt,f);
  t4=toc();
  xstart=[y0,ydot0];
  tic;yy5=dassl(xstart,t0,tt,res);
  t5=toc();
  zz=[t1,t2,t3,t4,t5]
  xbasc()
  plot2d(tt,yy1(7,:))
  plot2d(tt,yy2(7,:))
  plot2d(tt,yy3(7,:))
  plot2d(tt,yy4(7,:))
  plot2d(tt,yy5(8,:))
endfunction
```

The functions in `compode.sci` are loaded in Scilab, and to compare the different methods on an easy problem we enter

```
⟼ comparison(-0.1,2,-3,0.1)
  ans  =

!   0.318    0.197    0.348    0.525    0.533 !
```

We see that the methods were comparable, with the Adams method the fastest and `dassl` the fastest of the stiff methods. Adams is often fastest on problems that are not rapidly varying. rkf is second-best. Generally, Runge–Kutta methods like rkf take more work per time step than a multistep of the same oder such as adams. Also, the adams can go to higher order.

However, some problems, such as those in electrical engineering and some mechanical systems can be highly oscillatory. In other areas such as chemical reactions we can have very rapidly decaying terms. To examine what happens with a rapidly decaying component we enter

```
⟼ comparison(-0.1,2,-5000,0.1)
  ans  =

!   0.471    2.885    0.422    70.131    0.745 !
```

We see that the implicit methods default, stiff, and dassl are much faster. adams takes 7 times as long as stiff and rkf takes 30 times as long as stiff. The default takes longer than stiff since it starts as a nonstiff method and then has to switch. As is not unusual, dassl is comparable to stiff but sometimes takes a little longer. That is because dassl is designed for fully implicit DAEs and sometimes has to do a bit more work.

If a system is highly oscillatory, then some of the methods designed for stiff problems have to reduce their step size or order for stability reasons. To illustrate, we enter

```
⟼ comparison(-0.1,100,-3,0.1)
  ans  =

!   11.787    4.678    8.68    5.793    23.37 !
```

Whereas in the previous problem the methods of stiff type were clearly faster, we see here that the stiff methods were much slower. In the case of dassl it has some heuristics that look out for oscillatory systems and reduces the order to try to increase stability.

Sometimes simulation is embedded into design or optimization loops so that the same differential equations will be integrated many times. As the previous example shows, the computational time can vary greatly with the choice of the method. Accuracy can also vary, so that for complex problems one needs to do a careful examination of which integrator will be fastest and also deliver the desired accuracy.

Implicit Differential Equations

In some application areas the models are most naturally initially formulated as differential equations in the form

$$A(t,y)\dot{y} = g(t,y), \quad y(t_0) = y_0. \tag{3.21}$$

If $A(t,y)$ is not invertible for all t, y of interest, then (3.21) is a linear implicit differential algebraic equation. If $A(y,t)$ is invertible for all t, y of interest, we call (3.21) either a linearly implicit differential equation or an index-zero DAE. For this section we assume that $A(t,y)$ is an invertible matrix. One could, of course invert A and get the ODE

$$\dot{y} = A(t,y)^{-1}g(t,y) \quad y(t_0) = y_0. \tag{3.22}$$

However, this reformulation is not always desirable. The implicit system (3.21) can often be integrated more efficiently and reliably than (3.22). The numerical methods for solving (3.21) are similar to those we discuss later for DAEs of the form $F(t, y, \dot{y}) = 0$, which require a way to evaluate F given estimates of $y(t)$, $\dot{y}(t)$. Traditionally this function F is called a residual since iterative numerical methods are used, and when approximations are substituted in for y and $\dot{y}$, the evaluation of $F(t, y, \dot{y})$ tells us how far the approximations are from being solutions.

One Scilab simulation tool for (3.21) is impl. It is called by

```
y=impl([type],y0,ydot0,t0,t [,atol, [rtol]],res,adda [,jac])
```

Most of these variables have the same meaning as they do for ode, but there are some important new variables and there are some restrictions on the other variables. If type is present, it is either "adams" or "stiff". The algorithm requires an initial value of $\dot{y}(t_0)$ and this is provided in the vector ydot0.

The actual differential equation being integrated enters through the residual res, which is an external, that is, a function with specified syntax, or the name of a Fortran subroutine or a C function (character string) with specified calling sequence or a list.

If res is a function, its syntax must be r = res(t,y,ydot) and it must return the value of g(t,y)-A(t,y)*ydot.

In the implicit methods there are nonlinear equations that must be solved, and these require Jacobian information. The variable adda tells the integrator what A is. It also is useful within the numerical algorithm in setting up other Jacobians. If adda is a function, its syntax must be r = adda(t,y,p) and it must return r=A(t,y)+p, where p is a matrix to be added to A(t,y). As usual, if adda is a character string, it refers to the name of a Fortran subroutine or a C function.

The optional input jac is also an external. If jac is a function, it must have the syntax j = jac(t,y,ydot) and it must return the Jacobian of

$$r= g(t,y) -A(t,y)*ydot$$

with respect to y.

If a differential equation is fully implicit, that is, $F(t, y, \dot{y}) = 0$ with $\partial F / \partial \dot{y}$ nonsingular, then it is integrated using dassl as described in the section on differential algebraic equations.

Linear Systems

Since they occur so often in areas such as control engineering, there are a number of specialized functions for working with linear systems of the form

$$\dot{x} = Ax + Bu, \tag{3.23a}$$
$$y = Cx + Du, \tag{3.23b}$$

where A, B, C, D are matrices. System (3.23) is called the state-space form. If we let $x(0) = 0$, take the Laplace transform of (3.23), and let $\hat{z}(s)$ denote the Laplace transform of $z(t)$, we see that (3.23) can be written as

$$\hat{y}(s) = (D + C(sI - A)^{-1}B)\hat{u}(s) = H(s)u(s), \tag{3.24}$$

where H is a rational matrix function. That is, the entries of H are fractions of polynomials. Further manipulations would allow us to get

$$D(s)\hat{y}(s) = N(s)\hat{u}(s) = H(s)u(s), \tag{3.25}$$

where D, N are matrices of polynomials in s.

Scilab has functions for working with all three of these ways of describing linear systems. It also easily switches from one representation to another. Linear systems are defined by the `syslin` function. It may be called in three different ways:

```
[sl]=syslin(dom,A,B,C [,D [,x0] ])
[sl]=syslin(dom,N,D)
[sl]=syslin(dom,H)
```

The resulting `sl` is a tlist ("syslin" list) representing the linear system. `sl` can in turn be used in a number of different Scilab functions. `dom` is a character string (`'c'` , `'d'` , or `[]`) or a scalar. It defines the domain of the system. `'c'` is for a continuous-time system, `'d'` is for a discrete-time system, and a scalar `n` is for a sampled system with sampling period `n` (in seconds). A, B, C, D are the matrices of the state-space representation (3.23). x0 is the initial state. If x0 is not given, then x0=0. N, D, and H are polynomial and rational matrices respectively corresponding to (3.25) and (3.24).

If it is desired to recover the values of A, B, C, D, from `sl`, this is done by the `abcd` command,

$\longmapsto$ `[A,B,C,D]=abcd(sl)`

The scilab command `csim` provides a simulation using `ode` of the linear system (3.23) defined by `ls`. `csim` is called by

```
[y [,x]]=csim(u,t,sl,[x0 [,tol]])
```

u is a function, list, or string that defines the control u. t provides the times requested for the simulation values. t(1) is the initial time. y is a matrix of y values such that y(i)=[y(t(i)] for $i = 1, \ldots, n$. The optional output x is the corresponding values of the state x. tol is the two-vector [atol rtol] defining absolute and relative error tolerances for the ode solver. In addition to the usual choices for u such as a function, list, or vector of values, there are two special options. If u is the string "impuls", then an impulse response calculation is done. That is, u is a delta function, or unit impulse, at time $t = 0$. If u is "impuls", then it is assumed that sl is SISO (single input-single output), $D = 0$, and x0=0. u can also be "step" for a step response calculation. That is, $u = 1$ for $t > 0$. "step" has the same assumptions as "impuls".

Example 9. Consider the continuous linear system

$$\dot{x}_1 = -2x_1 + 7x_2 + u, \tag{3.26a}$$
$$\dot{x}_2 = 8x_1 - 2x_2, \tag{3.26b}$$
$$y = x_1 + x_2. \tag{3.26c}$$

When executed, the following file produces the output for a step input and an impulse input in two different windows.

```
A=[-2 7;-8 -2];B=[1;0];C=[1 1 ];D=0;
[sl]=syslin("c",A,B,C);
t=0:0.01:5;
[ys,xs]=csim("step",t,sl);          ← Finds step response of system
plot2d(t,ys);                        ← Fig. 3.3 (upper graph)
```

```
[yi,xi]=csim("impuls",t,sl);                    ← Finds impuls response
f1=scf(1);
xbasc();
plot2d(t,yi)                                     ← Fig. 3.3 (lower graph)
```

The graphs are given in Figure 3.3.

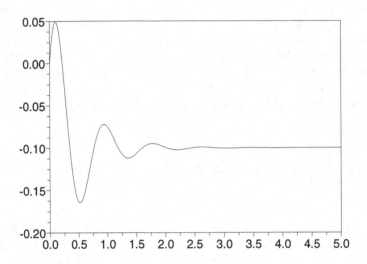

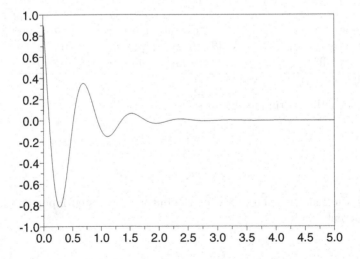

Figure 3.3. Output step response (above) and impulse response (below) for (3.26).

Root Finding

It is often desirable to simulate a differential equation up to the time something happens. In some cases the time is the quantity of interest, for example in an interception problem. In other cases the time will represent physical change in the dynamics such as when a tank is filling up and it starts to overflow. In all these cases we have a system (3.1) and we wish to integrate it until some quantity $g(t, y)$ is equal to 0. Scilab includes root finding software for both ODEs and DAEs. We focus on the ODE case here. The Scilab function ode_root is called by

```
[y,rd [,w,iw]]=ode("root",y0,t0,t [,rtol [,atol]],f [,jac],ng,g [,w,iw])
```

All of these variables have the same meaning as they do for ode, except that there are two new input variables g and ng, and a new output variable rd. g is an an external that defines $g(t, y)$, and ng is its size. The integration will proceed until the solution of (3.1) crosses the surface $g(t, y) = 0$. If g is a function, then the syntax should be z=g(t,y). The function g can be vector-valued. In that case the integration stops as soon as any component of g is zero. The output rd is a $1 \times k$ matrix. The first entry of rd contains the stopping time. Other entries indicate which components of g have changed sign. $k > 2$ indicates that more than one surface has been simultaneously traversed.

It is worth noting that ode_root does not just stop when it crosses the surface $g(t, y) = 0$. Rather, once it has crossed the surface, the algorithm iteratively reduces the last step in order to try to end on the surface up to the specified tolerances. This is an important feature.

When the surface-crossing represents a change in the differential equation, one could try to describe this situation using a single function with if statements. However, then there is the problem that the numerical simulation will step across the event and make the change at a later time. Often it is important to have the change occur at the correct time, and this requires the integrator searching out exactly when the crossing occurs.

Example 10. We suppose that we have a 1,000 cubic meter rectangular tank containing water that flows out a hole at the bottom. The tank is 10 meters high and initially full. When the tank is three quarters empty, a sensor causes water to be pumped into the tank at a constant rate. We are interested in the amount of water in the tank, which we denote by V. This may be modeled by $\dot{V} = -c_1\sqrt{V}$, $V(0) = 1,000$ until $V(t) = 250$, and then the differential equation is $\dot{V} = -c_1\sqrt{V} + c_2$, where c_1, c_2 are two positive constants determined by the size of the hole and the pumping rate. We are interested in the first 600 time units or until the tank fills back up, whichever occurs first. Listed below is a file tank.sci, which performs the simulation. It sets itask to 2 so that all mesh points are output. The if-then construction is built to cover the case in which t_f occurs before the surface is reached, which does not happen for the particular flow rates and final time chosen. This implementation also stops the simulation if the tank fills back up. The simulation is obtained by executing the next script, and the simulation graph is in Figure 3.4.

```
function z=f1(t,V)                          ← tank draining
  z=-0.3*sqrt(V);
endfunction

function z=f2(t,V)                          ← tank draining with pumping in
  z=-0.3*sqrt(V) +8.0;
```

```
endfunction

function z=g(t,V)                                              ← surface
  z=250-V;
endfunction

function z=h(t,V)                                       ← end of integration
  z=1000-V;
endfunction

%ODEOPTIONS=[2,0,0,%inf,0,2,10000,12,5,0,-1,-1];    ← set itask to 2, increase maxstep
tf=600;
[vsol,rd]=ode('root',1000,0,tf,f1,1,g);
m=size(vsol);
 if rd(1)< tf  then
   [vsol2,rd2]=ode('root',vsol(2,m(2)),rd(1),tf,f2,1,h);
   sol=[vsol,vsol2];
 else sol=vsol;
 end
xbasc();
plot2d(sol(1,:),sol(2,:));                                     ← Fig. 3.4
```

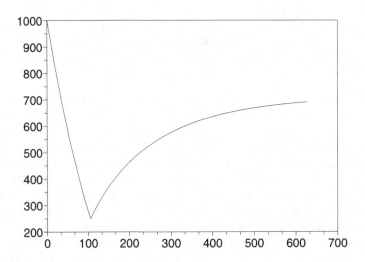

Figure 3.4. Simulation of Example 10 using ode_root.

3.2.2 Boundary Value Problems

A number of methods have been developed for solving BVP. The shooting methods take the given initial information and guess the rest of the initial information. They then adjust this initial guess by integrating over the full interval and use how far they miss the terminal boundary condition to correct the initial guess. Shooting methods are easy to program and can solve some problems, but they are not reliable for problems with long intervals or that are stiff in some sense. A more sophisticated approach is multiple shooting, which breaks the time interval into several subintervals and shoots over these.

The third approach is to discretize the differential equation using some numerical discretization, and then solve the large discrete system that results. For example, if one used Euler's method with a fixed step of h as the discretization, then the BVP (3.5) would become the nonlinear system of equations

$$x_{i+1} - x_i - hf(t_0 + ih, x_i) = 0, \quad i = 0, \ldots, N - 1, \tag{3.27a}$$
$$B(x_0, x_N) = 0. \tag{3.27b}$$

Of course, modern computer programs do not use Euler's method, but (3.27) makes the key point that what must be solved is a large system of equations, and these equations are nonlinear if the differential equation or the boundary conditions are nonlinear. This means that the BVP solver must deal with all the usual problems of numerically solving a nonlinear system including having a good enough initial guess and needing some sort of Jacobian information.

Scilab provides one simulation tool for BVP, which is bvode. It uses the COLNEW code, which is an update of the COLSYS code [6, 7, 8, 23]. Unlike the situation with ode and dassl, which requires the differential equations to be in first-order form, bvode allows the equations to be of higher order. bvode also allows for boundary conditions at several times ζ_j. It is traditional in parts of the BVP literature to use x instead of t for the independent variable. We will now switch to that notation.

bvode assumes that the BVP is in the form

$$\frac{d^{m_i} u_i}{dx} = f_i\left(x, u(x), \frac{du}{dx}, \ldots, \frac{d^{m_i-1}u}{dx^{m_i-1}}\right), \ 1 \leq i \leq n_c, \tag{3.28a}$$
$$g_j\left(\zeta_j, u(\zeta_j), \ldots, \frac{d^{m_*}u}{dx^{m_*}}\right) = 0, \ j = 1, \ldots, m_*, \tag{3.28b}$$

where ζ_j are the x values where the boundary conditions are to hold and $a_L \leq x \leq a_R$. In order to keep the notation down, let

$$m_* = m_1 + m_2 + \cdots + m_{n_c}, \tag{3.29}$$
$$z(u) = \left[u, \frac{du}{dx}, \ldots, \frac{d^{m_*}u}{dx^{m_*}}\right]. \tag{3.30}$$

Then (3.28) becomes

$$\frac{d^{m_i} u_i}{dx} = f_i(x, z(u(x)), \ 1 \leq i \leq n_c, \ a_L \leq x \leq a_R, \tag{3.31a}$$
$$g_j(\zeta_j, z(u(\zeta_j)) = 0, \ j = 1, \ldots, m_*. \tag{3.31b}$$

bvode starts with an initial mesh for the discretization and solves the resulting nonlinear system. It then iteratively refines the mesh until either it estimates that the requested accuracy has been attained or it reaches the maximum allowable mesh. The call for bvode takes the form

```
[z]=bvode(points,ncomp,m,aleft,aright,zeta,ipar,ltol,tol,fixpnt,fsub1,...
        dfsub1,gsub1,dgsub1,guess1)
```

The parameters in the bvode call have the following meanings:

z: The solution of the ode evaluated on the mesh given by `points`.

`points`: An array that gives the points for which we want the solution.

`ncomp`: The number of differential equations. It must satisfy `ncomp` $\leq$ 20.

`m`: A vector of size `ncomp` . The entry `m(j)` gives the order of the jth differential equation. The orders must satisfy `m(i)` $\leq$ 4.

`aleft`: The left end of the interval on which u is defined.

`aright`: The right end of the interval on which u is defined.

`zeta`: Entry `zeta(j)` gives the jth side condition point (boundary point). The entries must be ordered so that `zeta(j)` $\leq$`zeta(j+1)`. All the side (boundary) conditions must be at mesh points in all meshes used. In particular, they must be part of the initial mesh. Note the description of `ipar(11)` and `fixpnt` below.

`ipar`: An integer array with dimension at least 11. A list of the parameters in `ipar` and their meaning follows. Note that some parameters are renamed in `bvode`. These new names are given in parentheses.

`ipar(1)`: 0 if the problem is linear and 1 if the problem is nonlinear.

`ipar(2)`: The number of collocation points per subinterval ($= k$) where $\max m(i) \leq k \leq 7$. If `ipar(2)`$=0$, then `bvode` sets $k = \max(\max_i m(i) + 1, 5 - \max_i m(i))$.

`ipar(3)`: The number of subintervals in the initial mesh ($= n$). If `ipar(3)` $= 0$, then `bvode` sets `n = 5`.

`ipar(4)`: The number of solution and derivative tolerances. (= `ntol`) It is required that $0 <$ `ntol` $\leq$ `mstar`.

`ipar(5)`: The dimension of `fspace` (= `ndimf`). This is a real work array. Its size provides a constraint on `nmax`. Choose `ipar(5)` according to the formula `ipar(5)` $\geq$ `nmax*nsizef`, where `nsizef=4+3*mstar +(5+kd)*kdm+(2*mstar-nrec)*2*mstar`.

`ipar(6)`: The dimension of `ispace` (= `ndimi`), an integer work array. Its size provides a constraint on `nmax`, the maximum number of subintervals. Choose `ipar(6)` according to the formula `ipar(6)>=nmax*nsizei` where `nsizei=3 + kdm` with `kdm=kd+mstar`, `kd=k*ncomp`, and `nrec` = the number of right end boundary conditions.

`ipar(7)`: Provides output control (= `iprint`). It should be set = -1 for full diagnostic printout, = 0 for selected printout, and = 1 for no printout.

`ipar(8)`: This is (= `iread`). Setting it = 0 causes `bvode` to generate a uniform initial mesh. It should be = 1 if the initial mesh is provided by the user. Other values are not yet implemented in Scilab. The initial mesh is defined in `fspace`. The mesh will occupy `fspace(1)`,...,`fspace(n+1)`. the user needs to supply only the interior mesh points `fspace(j)` = `x(j)`, for $j = 2, \ldots, n$. If `ipar(8)`= 2, then the initial mesh is supplied by the user as with `ipar(8)=1`, but no adaptive mesh selection is to be done.

`ipar(9)`: This is (= `iguess`). It should be set = 0 if no initial guess for the solution is provided, = 1 if an initial guess is provided by the user in the subroutine `guess`, = 2 if an initial mesh and approximate solution coefficients are provided by the user in `fspace`. The former and new mesh are the same. It should be set = 3 if a former mesh and approximate solution coefficients are provided by the user in `fspace` but the new mesh is to be taken twice as coarse (that is, every second point from the former mesh), and = 4 if in addition to a former initial mesh and approximate solution coefficients, a new mesh is provided in `fspace` as well. See the description of `output` for further details on `iguess` = 2, 3, and 4.

`ipar(10)`: This is = 0 if the problem is regular. It is set = 1 if the first relaxation factor is =`rstart` and the nonlinear iteration does not rely on past convergence. Use this value for an extra-sensitive nonlinear problem only. Set = 2 if you want the code to return immediately upon (a) two successive nonconvergences or (b) after obtaining an error estimate for the first time.

ipar(11): This is the number of fixed points in the mesh other than `aleft` and `aright`. (= nfxpnt, the dimension of `fixpnt`). The code requires that all side condition points other than `aleft` and `aright` (see description of `zeta`) be included as fixed points in `fixpnt`.

ltol: An array of dimension ipar(4). ltol(j) = 1 specifies that the jth tolerance in `tol` controls the error in the lth component of z(u). It is also required that $1 \leq$ ltol(1) < ltol(2) < $\cdots$ < ltol(ntol) $\leq$ mstar.

tol: An array of dimension ipar(4). tol(j) is the error tolerance on the ltol(j) component of z(u). Thus, the code attempts to satisfy for $1 \leq j \leq$ ntol on each subinterval the following bound $|z(v) - z(u)|_{ltol(j)} \leq$ tol(j)$|z(u)|_{ltol(j)} +$ tol(j), where v(x) is the approximate solution vector and the subscript denotes the tol(j) entry.

fixpnt: An array of dimension ipar(11). It contains the points, other than `aleft` and `aright`, that are to be included in every mesh.

externals: The functions `fsub`, `dfsub`, `gsub`, `dgsub`, `guess` are Scilab externals. The Fortran-coded function interface to `bvode` is specified in the file `fcol.f`.

fsub: The name of the subroutine for evaluating f(x,z(u(x))) with ncomp entries at a point x in the interval (aleft, aright). It should have the heading [f]=fsub(x,z) where f is the vector containing the value of f(x,z(u)) and z(u(x))=(z(1),...,z(mstar)).

dfsub: Name of the subroutine for evaluating the Jacobian of f(x,z(u)) at a point x. it should have the heading [df]=dfsub(x ,z) where z(u(x)) is defined as for `fsub` and df is an ncomp× mstar array. The entries should be df(i,j), which are $\partial f_i / \partial z_j$.

gsub: Name of the subroutine for evaluating the ith component of g(x,z(u(x))) = g(zeta(i),z(u(zeta(i)))) at a point x = zeta(i), where $1 \leq i \leq$ mstar. It should have the heading [g]=gsub(i,z). Note that in contrast to f in `fsub`, here only one value per call is returned in g.

dgsub: Name of the subroutine for evaluating the ith row of the Jacobian of g(x,u(x)). It should have the heading [dg]=dgsub(i,z).

guess: Name of subroutine to evaluate the initial approximation for z(u(x)) and for dmval(u(x)) = vector of the m(j)th derivatives of u(x). It should have the heading [z,dmval]= guess(x). Note that this subroutine is used only if ipar(9) = 1, and then all mstar components of z and ncomp components of dmval should be specified for any x in the interval (aleft, aright) .

Example 11. One source of boundary value problems is in the necessary conditions for optimal control problems. Other techniques are used for more complicated optimal control problems [10] but solving the necessary conditions works on many simpler problems.

To illustrate suppose that we have the nonlinear controlled system

$$\dot{y} = y^2 + v$$

and we want to choose the control v so that it will steer y from 2 at time 0 to -1 at time 10. Given this requirement, we wish to have v minimize the quadratic cost

$$J(y,v) = \int_0^{10} 10v^2 + y^2 dt.$$

This cost penalizes v and y, so that it encourages both the control and state to be small.

Then the necessary conditions [36] are found by taking the Hamiltonian

$$H = 10v^2 + y^2 + \lambda(y^2 + v)$$

and forming the boundary value DAE

$$\dot{y} = \frac{\partial H}{d\lambda}, \tag{3.32a}$$

$$-\dot{\lambda} = \frac{\partial H}{\partial y}, \tag{3.32b}$$

$$0 = \frac{\partial H}{\partial v}, \tag{3.32c}$$

$$y(0) = 2, \quad y(10) = -1, \tag{3.32d}$$

which is

$$\dot{y} = y^2 + v, \tag{3.33a}$$

$$\dot{\lambda} = -2y - 2\lambda y, \tag{3.33b}$$

$$0 = 20v + \lambda, \tag{3.33c}$$

$$y(0) = 2, \quad y(10) = -1. \tag{3.33d}$$

Using (3.33c) to eliminate v from (3.33a), we get the BVP

$$\dot{y} = y^2 - \lambda/20, \tag{3.34a}$$

$$\dot{\lambda} = -2y - 2\lambda y, \tag{3.34b}$$

$$y(0) = 2, \quad y(10) = -1. \tag{3.34c}$$

We will now solve the BVP using bvode. To simplify things, we shall use the defaults whenever possible. In particular, we will not provide Jacobians or an initial guess. The following script sets up all the needed values for the call. A dummy guess is included since the call requires something in the guess position even if not used. After the BVP is solved the optimal $y(t)$ and $v(t)$ are plotted on the same graph as shown in Figure 3.5. The solid line is $y(t)$ and the x-line curve plots $v(t)$. In this problem the origin is unstable. The control has to work harder to get the solution near zero at the start than it does at the end of the interval.

```
function w=f(x,z)
 w1=z(1)^2-z(2)/20;
 w2=-2*z(1)-2*z(1)*z(2)
 w=[w1;w2];
endfunction

function w=df(x,z)
 w=[2*z(1),-1/20;-2-2*z(2),-2*z(1)]
endfunction

function w=g(i,z)
  if i==1  then w=z(1)-2,
  else w=z(1)+1;
  end
endfunction

function w=dg(i,z)
 w=[1 0]
```

endfunction

```
tt=0:0.1:10;
ncomp=2;nrec=1;
m=[1 1];k=4;                                    ← k=4 since ipar(2)=0 below
nmax=200;                                       ← set the maximum number of subintervals

mstar=2;                                        ← begin computation for ipar(5), ipar(6)
kd=k*ncomp;
kdm=kd+mstar;
nsizei=kdm+3;
ip6=nmax*nsizei;
nsizef=4+3*mstar+(5+kd)*kdm+(2*mstar-nrec)*2*mstar;
ip5=nmax*nsizef;

zeta=[0 10];                                    ← aleft=0 and aright=10
ipar=[1 0 0 2, ip5, ip6,   0 0 0 0 0];
er=1.0d-3;
ltol=[1 1];
tol=[er,er];
```

```
function w=gs(x)
  w=[0,0;0,0]
endfunction
```

```
zz=bvode(tt,2,m,0,10,zeta,ipar,ltol,tol,0,f,df,g,dg,gs);
plot2d(tt,zz(1,:));
control=-(1/20)*zz(2,:);
plot2d(tt,control);
plot2d(tt,control,style=-1)                     ← Fig. 3.5
```

3.2.3 Difference Equations

In many ways the simulation of discrete systems is easier because there is no choice about the time step and no error is introduced by the approximation of derivatives. Any error is the usual function evaluation and roundoff error. In order to simulate a discrete system of the first-order form

$$y(k + 1) = f(k, y(k)), \quad y(k_0) = y_0, \tag{3.35}$$

we use ode_discrete, which is called by

⟼ y=ode("discrete",y0,k0,kvect,f)

Note that the initial time k0, is an integer and kvect, which contains the times at which we want the solution, is a vector of integers. kvect(1) must be greater than or equal to k0. As an illustration, suppose that we enter the scilab code

```
function z=f(t,y)
  z=-y/2 + sin(t/10)
endfunction
kvect=0:20;
```

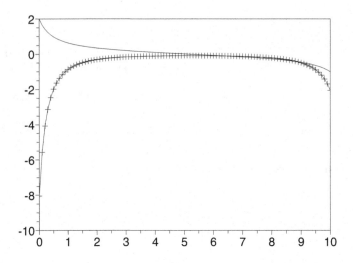

Figure 3.5. Optimal y (line) and control u (x-line) for Example 11.

```
y=ode("discrete",1,0,kvect,f);
plot2d(kvect,y)                                                    ← Fig. 3.6

xbasc()
plot2d2(kvect,y)                                                   ← Fig. 3.7
```

If we want to plot the graph, we might use the command `plot2d(kvect,y)`, which results in Figure 3.6. However, this graph is misleading since the actual solution exists only at integer values. A better option is to just plot either the points or to use `plotd2d2(kvect,y)` as in Figure 3.7.

Linear Systems

As with the continuous case, there are specialized programs for working with linear systems with real matrix coefficients in the form of

$$x(k+1) = Ax(k) + Bu(k), \qquad (3.36a)$$

$$y(k) = Cx(k) + Du(k). \qquad (3.36b)$$

If we just wish to simulate the state equation (3.36a), then we can use `ltitr`. It is called with either of

```
[X]=ltitr(A,B,U,[x0])
[xf,X]=ltitr(A,B,U,[x0])
```

With `ltitr` the initial value of k is understood to be zero. If the initial value x0 is not given, it is chosen as the zero vector. The inputs $u(i)$ are the columns of the U matrix, U=[u0,u1,...,un]. X is the matrix of outputs, so that X=[x(0),x(1),x(2),...,x(n)], and xf is the vector of the final state xf=X[n+1].

If the system is given by a transfer function, then we can use `rtitr` with calling syntax

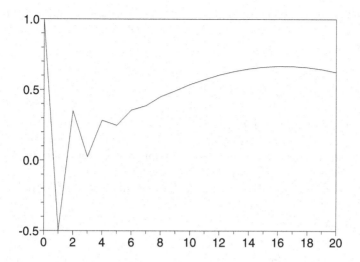

Figure 3.6. Graph produced by `plot2d(kvect,y)`.

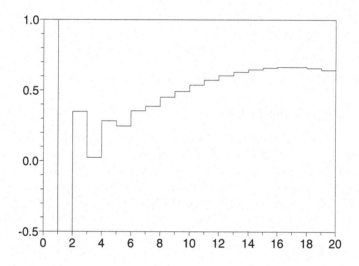

Figure 3.7. Graph produced by `plot2d2(kvect,y)`.

```
[y]=rtitr(Num,Den,u [,up,yp])
```

Here `Num` and `Den` are polynomial matrices of dimensions $n \times m$ and $n \times n$ that are the numerator and denominator of the transfer function $D^{-1}(z)N(z)$. If $D(z) = \sum_{i=0}^{d_1} D_i z^i$ and $N(z) = \sum_{i=0}^{d_2} N_i z^i$, then $D(z)y = N(z)u$ is interpreted as

$$\sum_{i=0}^{d_1} D_i y(t+i) = \sum_{i=0}^{d_2} N_i u(t+i)$$

for $t = 0, 1, \ldots$. It is assumed that D_{d_1} is nonsingular. The columns of `u` are the inputs of the system at $t = 0, 1, ..., T$. The outputs at $t = 0, 1, \ldots, T + d_1 - d_2$ are the columns of the matrix y, that is, `y=[y(0), y(1),...., y(T+d1-d2)]`. `up` and `yp` define the initial conditions for $-d_1 < t < 0$. The default values of `up` and `yp` are zero.

The time response, or simulation, of the linear system (3.36) can be done using `flts`, which is called by either of

```
[y [,x]]=flts(u,sl [,x0])
[y]=flts(u,sl [,past])
```

Here `sl` is a `syslin` list containing A, B, C, D or the information for the transfer function representation, `u` is the matrix (input vector), `x0` is the initial state with default value $= 0$, `past` is the matrix giving the past with default value $= 0$.

If one just wants the simulation of the output, then it is often easier to use `dsimul`, which provides state-space discrete-time simulation with syntax

```
y=dsimul(sl,u)
```

If `[A,B,C,D]=abcd(sl)` and `x0=sl('X0')`, then `dsimul` returns

```
y=C*ltitr(A,B,u,x0)+D*u.
```

With `dsimul` it is assumed that `sl` is in state-space form.

3.2.4 Differential Algebraic Equations

Before giving the simulation tools for DAEs it is helpful first to point out some of the requirements of these tools. With an ODE one had only to supply the initial value of $y(t_0)$. But a DAE needs information on both $y(t_0)$ and $\dot{y}(t_0)$ in order to uniquely determine the solution and get the integration started. Consider, for example, the implicit ODE $\tan(\dot{y}) = -y + g(t)$. This actually defines a family of differential equations $\dot{y} = \tan^{-1}(y + g) + n\pi$. Without a value for $\dot{y}$, the integrator would not be able to determine which family it should be integrating. There are a number of possibilities. A user might have a value of $\dot{y}$. Or one may have only an approximate value. Sometimes one has no idea at all. We return to this problem in Example 12.

The simulation tools in Scilab use backward differentiation formulas (BDF). For example, a backward Euler's method applied to (3.15) would take the form

$$F\left(t_{n+1}.y_{n+1}, \frac{y_{n+1} - y_n}{h}\right) = 0. \tag{3.37}$$

Given y_n, equation (3.37) must be solved by an iterative method. The Jacaobian with respect to y_{n+1} of (3.37) is $F_y + \frac{1}{h}F_{y'}$. This accounts for the different form that the input jac has in `dassl` below from what it had in some of the earlier simulation tools.

Scilab provides the DAE simulation tool `dassl`, which is the DASSL code developed by Linda Petzold [12]. DASSL integrates general nonlinear fully implicit index-one DAEs. It can also be used on implicit ODEs. DASSL is called by [r [,hd]]=dassl(x0,t0,t [,atol,[rtol]],res [,jac] [,info] [,hd])

If y is n-dimensional, then r will have $2n + 1$ rows. The first row is the returned values of t. The next n rows are the corresponding values of y, and the last n rows are the corresponding values of $\dot{y}$. info is a list that passes information on to DASSL that is specific to the solution of index-one DAEs. It also provides some of the information provided by %ODEOPTIONS for ode. It will be described more carefully below.

The inputs to dassl have the following meanings:

x0: This can can be either y0, in which case ydot0 is estimated by dassl with zero as its first estimate of ydot0, or x0 can be the matrix [y0, ydot0]. If ydot0 is provided, then F(t,y0,ydot0) must be equal to zero. If you know only an estimate of ydot0, then set info(7)=1.

t0: This is the initial time.

t: As with ode, the vector t gives the times that the solution is outputted. If you want the solution at every time step, then set info(2)=1.

atol, rtol: As with ode, atol and rtol are real scalars or column vectors of the same size as y giving the absolute and relative error tolerances requested of the solution.

res: This is an external as with impl. res computes the value of F(t,y,ydot). If res is a function, its calling sequence must be [r,ires]=res(t,y,ydot). res must return the residue r=g(t,y,ydot) and an error flag ires. ires = 0 if res succeeds in computing r. ires =-1 if the residue is not defined for the given values of (t,y,ydot). ires =-2 if the parameters are out of the admissible range.

jac: This is an external (function or list or string) that computes the value of the Jacobian dF/dy+cj*dF/dydot for a given value of the parameter cj. If jac is a function, its calling sequence must be r=jac(t,y,ydot,cj) and it must return

r=dF(t,y,ydot)/dy+cj*dF(t,y,ydot)/dydot

where cj is a real scalar.

info: This is a list that contains 7 elements. Its default value is

list([],0,[],[],[],0,0)

info(1): real scalar that gives the maximum time for which F is allowed to be evaluated or an empty matrix [] if no limits are imposed for time.

info(2): flag that indicates whether dassl returns its intermediate computed values (flag=1) or only the user-specified time point values (flag=0).

info(3): 2-component vector that gives the definition [ml,mu] of the band matrix computed by jac;
r(i - j + ml + mu + 1,j) = "dF(i)/dy(j)+cj*dF(i)/dydot(j)".
If jac returns a full matrix, set info(3)=[].

info(4): Real scalar that gives the maximum step size. Set info(4)=[] if no limitation.

info(5): Real scalar that gives the initial step size. Set info(4)=[] if not specified.

info(6): Set info(6)=1 if the solution is known to be nonnegative; otherwise set info(6)=0.

info(7): Set info(7)=1 if ydot0 is just an estimate of $y'(t_0)$. Set info(7)=0 if F(t0,y0,ydot0)=0.

hd: Real vector that allows for storing the dassl information and to resume integration.

r: This is a real matrix. Each column is the vector [t;x(t);xdot(t)], where t is the time for which the solution had been computed.

Example 12. As an illustration of the use of dassl and also to point out the importance of the full initial conditions, we consider the implicit ODE

$$\tan(\dot{y}) = -y + 10t\cos(3t), \quad y(0) = 0. \tag{3.38}$$

Notice that $n\pi$ is a consistent initial value of $\dot{y}(0)$ for any integer value of n. This reflects the fact that there are several solutions consistent with (3.38). When executed, the following script uses dassl to solve (3.38) for $\dot{y}(0) = 0$, $\dot{y}(0) = \pi$, and $\dot{y}(0) = 2\pi$ and then plot them on the same axis. The resulting figure is Figure 3.8.

```
x0=[0 0];x1=[0, %pi];x2=[0,2*%pi];
t0=0;
tt=0:0.1:10;
 function [r,ires]=res(t,y,ydot)
   r=tan(ydot)+y-10*t*cos(3*t)
   ires=0
 endfunction
r0=dassl(x0,t0,tt,res);
r1=dassl(x1,t0,tt,res);
r2=dassl(x2,t0,tt,res);
plot2d(r0(1,:),r0(2,:));
plot2d(r1(1,:),r1(2,:));
plot2d(r1(1,:),r1(2,:),style=-1);
plot2d(r2(1,:),r2(2,:));
plot2d(r2(1,:),r2(2,:),style=-3);                      ← Fig. 3.8
```

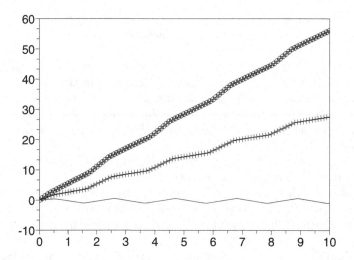

Figure 3.8. Simulation of (3.38) using dassl for three values of $\dot{y}(0)$.

The fact that implicit ODEs and fully implicit DAEs can have multiple nearby solutions for the same value of the state has important consequences for the design of these integrators since this behavior shows up in applications. The integrators must try to ensure that when making a step they do not jump to another solution. This is accomplished by being conservative in choosing the size of the step.

DAEs can also be formulated and solved in Scicos as described later. The integrator in ScicosLab Scicos is IDA from the SUNDIALS suite [21, 33]. IDA is a modified form of DASPK so it is a variable order backward differentiation code. IDA also has root finding capabilities.

DAEs and Root-Finding

Root-finding with DAEs is done with `dasrt`, which is `dassl` along with the ability to find where the solution of the DAE crosses a surface. The calling sequence for `dasrt` is the same as for `dassl` except for two inputs, which are the same as for `ode` using the type "root." The calling sequence of `dasrt` is

`[r,nn,[,hd]]=dasrt(x0,t0,t [,atol,[rtol]],res [,jac],ng, surf [,info] [,hd])`

The differences from the calling of `dassl` are `nn`, which is a vector with two entries `[times num]`. `times` is the value of the time at which the surface is crossed, and `num` is the number of the crossed surface. `surf` is an external (function or list or string) that computes the value of the column vector `surf(t,y)` with `ng` components. Each component defines a surface. If `surf` is a function, its calling sequence must be `surf(t,y)`.

3.2.5 Hybrid Systems

Many systems have both continuous-time and discrete-time components. The simulation tool `odedc` is designed to integrate some of these models. It is assumed that there are continuous variables y_c and discrete variables y_d that can change at discrete times t_k. y_d is considered as a function of continuous time by making it piecewise constant on the intervals $[t_k, t_{k+1}[$.

The system that is integrated then takes the form

$$\dot{y}_c(t) = f_0(t, y_c(t), y_d(t)), \ t \in [t_k, t_{k+1}[, \tag{3.39a}$$

$$y_d(t_{k+1}) = f_1(t, y_c(t_{k+1}), y_d(t_k)) \text{ at } t = t_{k+1}. \tag{3.39b}$$

In a given subinterval $[t_k, t_{k+1}]$, the software integrates (3.39a) up to time t_{k+1} using $y_d(t) = y_d(t_k)$. Then it resets the value of y_d using (3.39b) and then proceeds with integration at the next step. The parameters require that the t_k be uniformly spaced, but they do not require that the discrete and continuous systems start at the same time.

A special example of (3.39) is a sampled data system. In a sample data system the dynamics are given by a differential equation but the outputs occur at discrete times. A sampled data system could take the form

$$\dot{y}_c(t) = f_0(t, y_c(t), u(t)), \ t \in [t_k, t_{k+1}[, \tag{3.40a}$$

$$y_d(t_{k+1}) = f_1(t, y_c(t_{k+1}), u(t_k)) \text{ at } t = t_{k+1}, \tag{3.40b}$$

where u is a control function.

The calling sequence for `odedc` is

`yt=odedc(y0,nd,stdel,t0,t,f)`

Here the input terms of `odedc` have the following meaning:

y0: This is a real column vector of initial conditions for both the continuous and discrete variables, y0=[y0c;y0d].

nd: This is the number of components in y0d.

t0: This is the initial time for starting the simulation.

stdel: This is a real vector with one or two entries, stdel=[h, delta]. If there is one entry, then delta=0. h is $t_{k+1} - t_k$ for $k > 0$. That is, h is the length of the time interval between events. The variable delta = delay/h, where delay is the length of time after t_0 before the first discrete event happens.

t: This is instants where the solution is calculated and output.

f: This is an external. If it is a function, it must have the calling sequence

   ycd=f(t,yc,yd,flag).

This function must return the derivative of the vector yc if flag=0. That is, it evaluates the f_0 function in (3.39a). If flag=1, then f returns the update of yd. That is, it provides the function f_1 of (3.39b).

yp: This must be a vector with the same dimension as yc if flag=0. yp must be a vector with the same dimension as yd if flag=1.

y: This contains the values of the solution at times in t.

odedc can be called with the same optional parameters as the ode function, provided nd and stdel are given in the calling sequence as second and third parameters. In particular, integration flags and tolerances can be set. Optional parameters can be set by the odeoptions function.

As an example we consider the following example of a hybrid system:

$$\dot{y}_1 = -4y_2 + (z_k - 1)y_1 + 2, \quad y_1(0) = 1, \tag{3.41a}$$

$$\dot{y}_2 = 4y_1 + (z_k - 1)y_2, \quad y_2(0) = 1, \tag{3.41b}$$

$$z_{k+1} = -0.9z_k + 1.1\text{sgn}(y_1(t)), \quad z_0 = 0, \tag{3.41c}$$

where $t_1 = 3.3$ and $t_{k+1} - t_k = 3$ for $k > 0$ and we wish to simulate it on the time interval from 0 to 20.

We execute the following Scilab script:

```
function  ycd=f(t,yc,yd,flag)
  if flag==0  then
    ycd(1)= -4*yc(2)+(yd-1)*yc(1) + 2;
    ycd(2)=4*yc(1)+(yd-1)*yc(2);
  else
    ycd=-0.9*yd+ 1.1*sign(yc(1));
  end
endfunction

t0=0;y0=[1;1;0];nd=1;stdel=[3 1.1];
tt=0:0.1:20;
yt=odedc(y0,nd,stdel,t0,tt,f);
xbasc();
plot2d(tt',yt');                                    ← Fig. 3.9
```

The result is the plot of the three components in Fig. 3.9. The graph of yd is given by the square wave form. The two components of yc are the wiggly curves that change their behavior each time that yd changes.

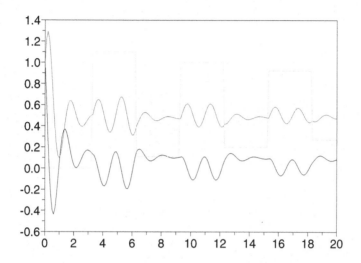

Figure 3.9. Simulation of (3.41) using odedc.

We conclude with an example of a sampled data system.

Example 13. Suppose that we have started with the linear system

$$\dot{x} = \begin{pmatrix} 0.1 & -1 \\ 1 & 0.1 \end{pmatrix} x + \begin{pmatrix} 1 \\ 0 \end{pmatrix} u. \tag{3.42}$$

This system is unstable. A feedback control

$$u = \begin{pmatrix} -2 & -1 \end{pmatrix} x \tag{3.43}$$

is designed to stabilize the system. The control is to be applied starting at time $t = 15$. Sometimes it is possible to measure the state at any time, but suppose that we are only able to determine the state every ρ time units starting at time $t = 5$. The following script when executed performs three simulations. The first is using the continuous version of the control. The second and third are with $\rho = 1$ and $\rho = 0.9$.

Figure 3.10 shows the effect of the continuous-time feedback. Once the feedback is applied at $t = 15$ it quickly stabilizes the system. If the sampling is rapid enough, the result is essentially the same as using the continuous feedback. However, as Figure 3.11 shows, once the sampling period is $\rho = 0.9$, the system is still stabilized, but it takes much longer for the state to go to zero. Once the sampling period reaches 1, the control no longer stabilizes the system, as shown in Figure 3.12.

```
function z=control(t,x)
```

```
   if t<15   then z=0,
   else z=-2*x(1)-x(2);
   end
endfunction

function xdot =f(t,x)
  xdot=[.1 -1;10.1]*x +[control(t,x);0]
endfunction

x0=[1;1];
tt=[0:0.1:40];
y=ode(x0,0,tt,f);
xbasc();
plot2d(tt,y(1,:));

xbasc()
xd0=[1;1];
xc0=[0;0];

 function ycd =fcd(t,yc,yd,flag)
   if flag==0  then ycd=[.1 -1;10.1]*yc +[control(t,yd);0],
   else ycd=yc;
   end
 endfunction

yy=odedc([xd0;xc0],2,[1,5],0,tt,fcd);
plot2d(tt,yy(1,:));

xbasc()
yy=odedc([xd0;xc0],2,[0.9,5/0.9],0,tt,fcd);
plot2d(tt,yy(1,:));
```

← Fig. 3.10

← Fig. 3.12

← Fig. 3.11

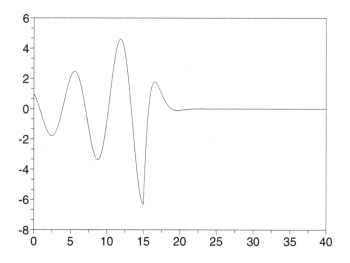

Figure 3.10. x_1 from simulation of Example 13 using continuous feedback.

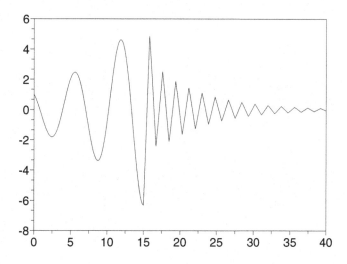

Figure 3.11. x_1 from simulation of Example 13 using sampled feedback with period $\rho = 0.95$.

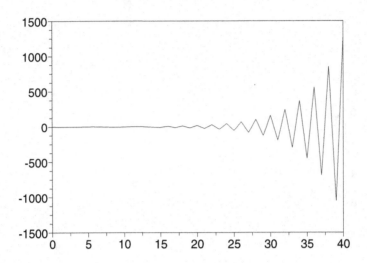

Figure 3.12. x_1 from simulation of Example 13 using sampled feedback with period $\rho = 1.0$.

4

Optimization

Various optimization problems play a fundamental role in modeling and simulation. This chapter will cover some of the optimization utilities available in Scilab. Section 4.1 will present some overview comments about optimization, useful in using the algorithms. Section 4.2 discusses the most general optimization utilities. There is a close relationship between optimization and solving equations in a least squares sense. This is examined in Section 4.4. One application of optimization is parameter fitting. Specialized utilities for this are given in Section 4.5. Finally, some specialized optimization utilities for linear and quadratic programming are given in Section 4.6.

4.1 Comments on Optimization and Solving Nonlinear Equations

In this chapter we discuss the problem of solving the minimization problem

$$\min_x f(x), \tag{4.1}$$

where f is a real-valued function of the vector variable x. Note that maximizing f is the same as minimizing $-f$, so that all of the utilities given here can also be used with maximization problems.

There are a number of characteristics of (4.1) that determine which, of any, of the available utilities will work. They also affect how the optional parameters should be set if a simple call does not work.

Constraints

The first question is whether there are any restrictions on x (4.1). if there are, the problem is said to be constrained. As far as Scilab is concerned there are three types of constraints.

Bound, or box, constraints require different entries of x to lie in specified intervals. For example, if x is 3 dimensional, we could have $-2 \leq x(1) \leq 3$ and $2 \leq x(3) \leq 4$.

Linear equality constraints take the form of $b'x - c = 0$ when b and x are written as column vectors. Linear inequality constraints take the form $b'x - c \leq 0$ when b, x are written as column vectors.

Finally there are the more general constraints of the form $g(x) = 0$. There are not currently any utilities in Scilab for solving general constrained optimal control problems. However, if the problem takes the form

S.L. Campbell et al., *Modeling and Simulation in Scilab/Scicos with ScicosLab 4.4*,
DOI 10.1007/978-1-4419-5527-2_4, © Springer Science+Business Media, LLC 2010

$$\min_x f(x), \tag{4.2a}$$
$$g(x) = 0, \tag{4.2b}$$

and f, g are differentiable, then the solution x^* of (4.2) satisfies the necessary conditions

$$f_x(x) + \lambda^T g_x(x) = 0, \tag{4.3a}$$
$$g(x) = 0, \tag{4.3b}$$

where g_x is the Jacobian of g, which is discussed below.

It is sometimes possible to solve (4.3) using `fsolve` as discussed later in this chapter.

The next question is whether f is a differentiable function and whether we know what its gradient is. If f is differentiable, then often some type of iterative method is used. Inside the iterative method at each iteration there is the determination of a search direction and a decision on how far to move in that direction. Then there are questions of how long the iteration will continue. Common choices are until f does not change much, until the gradient of f is small, or when the allowed number of iterations has been used. The last criterion can be important if function evaluations are computationally expensive.

These types of iterative methods, when they find a minimum, usually find local minima. That is, they find where f is smaller than at nearby values. Such local minima may or may not be a global minimum, that is, the place where f is smallest for all possible x.

A different type of algorithm is sometimes called a "dart-throwing algorithm." These algorithms vary in philosophy, but basically they continue to evaluate throughout the region but take more values where the function appears small. While usually much slower than iterative methods when both can be used, the dart-throwing algorithms are more apt to find a global minimum and multiple minimum values.

Nonlinear Equation Solving

A nonlinear equation takes the form

$$f(x) = 0. \tag{4.4}$$

In the simplest case this represents n equations in n unknowns. Such equations are also solved by iterative methods. A key role is played by the Jacobian matrix $J(x)$. If f has m entries and x has n entries, then

$$J(x) = \begin{pmatrix} \frac{\partial f_1}{\partial x_1} & \cdots & \frac{\partial f_1}{\partial x_n} \\ \vdots & \ddots & \vdots \\ \frac{\partial f_m}{\partial x_1} & \cdots & \frac{\partial f_m}{\partial x_n} \end{pmatrix}.$$

When $m = n$, and $J(x)$ is invertible, the simplest iterative method is Newton's method, which is

$$x_{j+1} = x_j - J(x_j)^{-1} f(x_j).$$

Actual solvers utilize a number of methods to increase the region of convergence and to provide estimates of J.

4.2 General Optimization

The general optimization utility in Scilab is `optim`. This algorithm allows the user to have some control of the optimization process. It can accept bound constraints. `optim` needs to know the gradient. If the gradient is unknown, or too complicated to code, then a utility is provided to enable optim to proceed using only the function to be minimized. This utility is NDcost, which stands for numerical differentiation of the cost. It approximates the gradient using finite differences.

There are additional software programs whose distribution is maintained by the developer but for which there exist Scilab interfaces. For example, a toolbox providing an interface for the general optimization code can be found at http://www.scilab.org/ under contributions, and then downloads, in the category "Optimization Tools." In particular, there is an interface to FSQP that can handle more general types of constraints than optim can.

We shall first present the simplified use of optim and NDcost. We will then discuss some of the more general options. The simplified call is

$\longmapsto$ [f,xopt]=optim(costf,x0)

Here x0 is the initial guess for where the minimum occurs, f is the optimum value, and xopt is where this optimum value occurs. costf is a function that provides the value of the function to be minimized, the value of its gradient, and a variable that is used by the optimization routine that we will explain later. If costf is a Scilab function, it takes the form

$\longmapsto$ [f,g,ind]=costf(x,ind)

Of course, for a more complicated problem we may not be able to, or prefer not to, provide the gradient. Then we can use the NDcost function. In its simplest form the call to optim then takes the form

$\longmapsto$ [f,xopt]=optim(list(NDcost,myf),x0)

In using optim and providing the gradient, the initial guess x0 can be either a row or column and then the xopt will be the same. However, in using NDcost, the initial guess must be a column vector.

Suppose that we wished to minimize the function

$$f(x, y, z) = (x - z)^2 + 3(x + y + z - 1)^2 + (x - z + 1)^2. \tag{4.5}$$

Its gradient is

$$\nabla f = \left[\frac{\partial f}{\partial x}, \frac{\partial f}{\partial y}, \frac{\partial f}{\partial z} \right]$$
$$= [2(x - z) + 6(x + y + z - 1) + 2(x - z + 1), 6(x + y + z - 1),$$
$$-2(x - z + 1) + 6(x + y + z - 1)], \tag{4.6}$$

and we take an initial guess of x0 = [0, 0, 0]. The following script solves using the gradient and then solves the problem using optim.

```
function  z= myf(x)
 z=(x(1)-x(3))^2+3*(x(1)+x(2)+x(3)-1)^2+(x(1)-x(3)+1)^2
endfunction

function z=myg(x)
 xs=x(1)+x(2)+x(3)-1;
 z=[2*(x(1)-x(3))+6*xs+2*(x(1)-x(3)+1),  6*xs,...
    -2*(x(1)-x(3))+6*xs-2*(x(1)-x(3)+1)]
```

endfunction

```
function [f,g,ind]=costf(x,ind)
 f=myf(x);g=myg(x);
```
endfunction

```
x0=[0 0 0];                               ← initial condition
[fopt,xopt]=optim(costf,x0);              ← x0 a row vector
[fopt,xopt]=optim(costf,x0');             ← x0 a column vector
[fopt,xopt,gopt]=optim(list(NDcost,myf),x0');   ← x0 must be a column
```

When executed, after listing the function definitions, the output gives the results:

```
⟼ x0=[0 0 0];                            ← initial condition
⟼ [fopt,xopt]=optim(costf,x0)            ← x0 a row vector
 xopt  =

!   0.0833333    0.3333333    0.5833333 !
 fopt  =

    0.5

⟼ [fopt,xopt]=optim(costf,x0')           ← x0 a column vector
 xopt  =

!   0.0833333 !
!   0.3333333 !
!   0.5833333 !
 fopt  =

    0.5

⟼ [fopt,xopt,gopt]=optim(list(NDcost,myf),x0')    ← x0 must be a column
 gopt  =

!   0.        !
!   0.        !
! - 1.833D-11 !
 xopt  =

!   0.0833333 !
!   0.3333333 !
!   0.5833333 !
 fopt  =

    0.5
```

In more complicated problems the user often has to set limits on how many iterations are taken in order to control the time of computation. In these cases the value of the gradient is very useful information, since if we are at a local optimum of an unconstrained smooth problem, then the gradient should be zero. Note that in the NDcost we get a numerical estimate of the gradient and it is close to zero. Requesting the value of the gradient at the optimal point is a general option with optim, as we will now show.

For more complicated problems, the user will want to use some of the options of `optim`, which we will now describe. The full calling format is

```
[f,[xopt,[gradopt,[work]]]]=optim(costf,[contr],x0,['algo'],....
            [df0,[mem]],[work],[stop],['in'],[imp=iflag])
```

These parameters are as follows. Using some of them requires knowledge about the iterative numerical methods used.

- `costf`: An external, Scilab function, list, or string (`costf` is the cost function: see its calling sequence below (Scilab, C, or Fortran)). `x0`: real vector (initial value of variable to be minimized).
- `f`: Value of optimal cost.
- `xopt`: Best value of `x` found.
- `contr`: The three entries `'b'`,`binf`,`bsup`. `binf` and `bsup` are real vectors with the same dimension as `x0`. `binf` and `bsup` are lower and upper bounds on `x`.
- `"algo"`: Determines the algorithm. `'qn'` is quasi-Newton, which is the default. `'gc'` is conjugate gradient. `'nd'` stands for nondifferentiable. Use this if there are points where the cost function does not have a derivative. An example is `abs`, which is not differentiable at zero. Note that `'nd'` does not accept bounds on `x`.
- `df0`: Real scalar giving the guessed decrease of `f` at first iteration. `df0=1` is the default value. It affects the efficiency of the line-search algorithm.
- `mem`: An integer giving number of variables used to approximate the Hessian when `algo='gc'` or `'nd'`. Default value is around 6.
- `stop`: A sequence of optional parameters controlling the convergence of the algorithm. `stop= 'ar',nap, [iter [,epsg [,epsf [,epsx]]]]`, where
  - `"ar"`: Reserved keyword for stopping rule selection defined as follows:
  - `nap`: Maximum number of calls to `costf` allowed.
  - `iter`: Maximum number of iterations allowed.
  - `epsg`: Threshold on gradient norm.
  - `epsf`: Threshold controlling decreasing of `f`.
  - `epsx`: Threshold controlling variation of `x`. This vector (possibly matrix) of same size as `x0`, can be used to scale `x`.
- `"in"`: Reserved keyword for initialization of parameters used when `costf` is given as a C or Fortran routine.
- `"imp=iflag"`: Named argument used to set the trace mode. If `iflag=0`, then nothing except errors are reported. If `iflag=1`, then there are initial and final reports. If `iflag=2`, then there is an additional report per iteration. `iflag>2` adds reports on linear search. Most of these reports are written on the Scilab standard output, so if there are many iterations involving vectors, there could be a lot of output.
- `gradopt`: Gradient of `costf` at `xopt`.
- `work`: Working array for hot restart for quasi-Newton method. This array is automatically initialized by `optim` when `optim` is invoked. It can be used as an input parameter to speedup the calculations.

The `costf` format is described above. The variable `ind` is used by `optim`. On output, `ind<0` means that `f` cannot be evaluated at `x` and `ind=0` interrupts the optimization. If `costf` is a character string, it refers to the name of a C or Fortran routine, which must be linked to Scilab. The generic calling sequence for the C or Fortran routine is the function `costf(ind,n,x,f,g,ti,tr,td)`. Details on these additional variables are given in the on-line help.

4.3 Solving Nonlinear Equations

The theory and algorithms for solving a nonlinear system of equations

$$\text{fct}(\text{x})=0 \tag{4.7}$$

intertwines in a number of ways with those for optimization. Assuming that f is differentiable, the primary Scilab utility is `fsolve`. It has the calling sequence

`[x [,v [,info]]]=fsolve(x0,fct [,fjac] [,tol])`

and uses the Powell hybrid method. `fsolve` is based on the package MINPACK [22].

The simplest calls need only

- `x0`: Real vector that is the initial value of the function argument.
- `fct`: External providing function in (4.7)

The optional parameters are

- `fjac`: External providing the Jacobian of `fct`.
- `tol`: Real scalar, which is a precision tolerance. Termination occurs when the algorithm estimates that the relative error between x and the solution is at most `tol`. Default value is `tol=1.d-10`.
- `x`: Final value of function argument. Estimated solution of (4.7).
- `v`: Value of function `fct` at x. Should be close to zero if a solution has been found.
- `info`: Indicates why termination occurred.
  - 0: improper input parameters.
  - 1: algorithm estimates that the relative error between x and the solution is at most `tol`.
  - 2: number of allowed calls to `fct` reached.
  - 3: `tol` is too small. No further improvement in the approximate solution x is possible.
  - 4: iteration is not making good progress.

The simplest calling sequence for `fct` is `[v]=fct(x)`. If `fct` is a character string, it refers to a C or Fortran routine, which must be linked to Scilab. Fortran calling sequence must be

```
fct(n,x,v,iflag)
integer n,iflag
double precision x(n),v(n)
```

and the C Calling sequence must be

```
fct(int *n, double x[],double v[],int *iflag)
```

As an illustration of using `fsolve` we consider the problem of

$$\min f(x,y,z) = (x-z)^2 + 3(x+y+z-1)^2 + (x-z+1)^2, \tag{4.8}$$

where

$$g(x,y,z) = 1 - x - 3y - z^2 = 0. \tag{4.9}$$

As noted earlier, the necessary condition for the minimum of this problem is that it be a solution of

$$\nabla f + \lambda \nabla g = 0, \tag{4.10}$$

$$g = 0. \tag{4.11}$$

If we execute the next script:

```
⟼  function z=fct(x)
⟼    xs=x(1)+x(2)+x(3)-1;
⟼    w1=[2*(x(1)-x(3))+6*xs+2*(x(1)-x(3)+1),6*xs,...
⟼        -2*(x(1)-x(3))+6*xs-2*(x(1)-x(3)+1)]
⟼    w2=[-1 -3 -2*x(3)]
⟼    z=[w1'+x(4)*w2';1-x(1)-3*x(2)-x(3)^2]
⟼  endfunction

⟼ x0=[0 0 0 0];                                    ← initial condition
⟼ [x,v]=fsolve(x0,fct);
```

we get

```
⟼ v                                               ← value of function at x
  ans  =

   1.0D-16 *

!   0.0016752    0.0050255    0.0023359  - 1.6653345 !

⟼ x                                                ← solution
  ans  =

!   0.1972244    0.1055513    0.6972244  - 1.675D-19 !
```

Since v= 0, we have found a solution given by x.

4.4 Nonlinear Least Squares

Suppose that $y = f(x)$ is a function from the space of n-dimensional vectors x to that of m-dimensional vectors y. Sometimes it may not be possible to solve $f(x) = 0$. The best that we can hope for is to find an x that makes $f(x)$ as small as possible. That is, you want x that minimizes $\|f(x)\|^2 = f(x)'f(x) = \sum_{i=1}^{m} f_i(x)^2$. Such a solution is called a least squares solution. Notice that this allows for $m > n$. That is, there are more equations than there are unknowns. This situation is important for a number of applications including parameter fitting. Scilab provides two utilities for nonlinear least squares. They are leastsq and lsqrsolve.

leastsq

The calling format of leastsq is much like that of optim except that we give the function f and optionally its Jacobian. The algorithm then computes the cost function $f'f$ and its gradient. The only other difference is that the parameters are in a different order than with optim. Thus the short calling format is

```
[f,xopt]=leastsq([imp,] fun [,Dfun],x0)
```

The complete format is

```
[f,[xopt,[gradopt]]]=leastsq(fun [,Dfun],[contr],x0,['algo'],...
                    [df0,[mem]],[stop],['in'])
```

Here `fun` is an external, i.e., a Scilab function or string defining the least squares problem.

Functions specialized for parameter fitting will be given shortly, but to illustrate the use of `leastsq`, suppose that we have the data points

$$\{(0,0),(0,1),(1,1,),(2,1.5),(2,2)\}$$

and we want to find parameters a, b, c such that $y = ae^{bt} + c$ fits these data as well as possible. If the unknown parameters are called $p = (a,b,c)$ and the data points (t_i, y_i), then we can solve in the least squares sense the five equations given by $y_i - p_1 e^{p_2 t_i} - p_3 = 0$. When executed, the following Scilab script finds p and plots in Figure 4.1 both the data points and the curve.

```
DAT=[0 0;0 1;1 1;2 1.5; 2 2];

function z=fun(p)
  DAT=[0 0;0 1;1 1;2 1.5; 2 2]
  z=DAT(:,2)-p(1)*exp(p(2)*DAT(:,1))-p(3)
endfunction

p0=[0 0 0];                                          ← initial value

[ff,p]=leastsq(fun,p0)                               ← ff value of the function at p
p  =

!   1.0000000    0.4054651   - 0.5000000 !
ff  =

    0.625

xbasc();
plot2d(DAT(:,1),DAT(:,2),-3,rect=[-0.5,-0.5,2.5,2.5])
t=linspace(-0.5,2.5,50);
et=p(1)*exp(p(2)*t)+p(3);
plot2d(t,et,rect=[-0.5,-0.5,2.5,2.5])                ← Fig. 4.1
```

lsqrsolve

This program minimizes the sum of the squares of nonlinear functions, using the Levenberg–Marquardt algorithm. The long and short forms of the calling sequence are

```
[x [,v [,info]]]=lsqrsolve(x0,fct,m [,stop [,diag]])
```

```
[x [,v [,info]]]=lsqrsolve(x0,fct,m ,fjac [,stop [,diag]])
```

depending on whether the Jacobian is supplied. Here

- `x0`: Real vector that is initial value of functions argument.
- `fct`: An external that defines the equations.
- `m`: Integer that is the number of functions.
- `fjac`: External that is the Jacobian of `fct`.

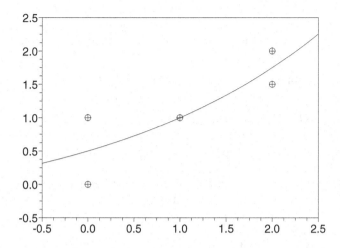

Figure 4.1. Data points and curve found by `leastsq`.

- `stop`: Optional vector `[ftol,xtol,gtol,maxfev,epsfcn,factor]` whose default value is `[1.d-8,1.d-8,1.d-5,1000,0,100]`. Controls the iterations.
  - `ftol`: A positive real number. Termination occurs when both the actual and predicted relative reductions in the sum of squares are at most `ftol`. `ftol` measures the relative error desired in the sum of squares.
  - `xtol`: A positive real number. Termination occurs when the relative error between two consecutive iterates is at most `xtol`. Thus `xtol` measures the relative error desired in the approximate solution.
  - `gtol`: A nonnegative input variable. Termination occurs when the cosine of the angle between `fct(x)` and any column of the Jacobian is at most `gtol` in absolute value. `gtol` measures the orthogonality desired between the function vector and the columns of the Jacobian.
  - `maxfev`: A positive integer. Termination occurs when the number of calls to `fcn` is at least `maxfev` by the end of an iteration.
  - `epsfcn`: A positive real number, used in determining a suitable step length for the forward-difference approximation. This approximation assumes that the relative errors in the functions are of order `epsfcn`. if `epsfcn` is less than the machine precision, it is assumed that the relative errors in the functions are on the order of the machine precision.
  - `factor`: A positive real number, used in determining the initial step bound. This bound is set to the product of `factor` and the Euclidean norm of `diag*x` if nonzero, or else to `factor` itself. In most cases factor should lie in the interval $(0.1, 100)$, and 100 is a generally recommended value.
- `diag`: This is an array of length `n`. `diag` must contain positive entries that serve as multiplicative scale factors for the variables.
- `x`: Final value of function argument, which is the estimated solution.
- `v`: Value of functions at `x`, also called the residual.
- `info`: Indicates cause of termination.
  - 0: improper input parameters.

– 1: algorithm estimates that the relative error between x and the solution is at most tol.
– 2: maximum number of calls to fcn has been reached.
– 3: tol is too small. No further improvement in the approximate solution x is possible.
– 4: iteration is not making good progress.

The functions describing the equations fct and their Jacobian fjac are of the same form. Letting fn stand for either fct or fjac, the allowable formats are

- A Scilab function whose calling sequence is v=fn(x,m) given x and m.
- A character string that refers to a C or Fortran routine that must be linked to Scilab. Should return −1 in iflag to stop the algorithm if the function or Jacobian could not be evaluated
  – Fortran calling sequence should be fct(m,n,x,v,iflag), where m, n, iflag are integers, x is a double-precision vector of size n and v a double-precision vector of size m.
  – C calling sequence should be
    fct(int *m, int *n, double x[],double v[],int *iflag)

Suppose that we wish not to use leastsq on our previous example, but wish to use lsqrsolve. The format of the function is slightly different due to the parameter m, but we can reuse our previous function by using the following script.

```
DAT=[0 0;0 1;1 1;2 1.5; 2 2];

 function z=fun(p)
   DAT=[0 0;0 1;1 1;2 1.5; 2 2]
   z=DAT(:,2)-p(1)*exp(p(2)*DAT(:,1))-p(3)
 endfunction

p0=[0 0 0];

 function z=fct(p,m)
   z=fun(p)
 endfunction

[p,v]=lsqrsolve(p0,fct,5);

xbasc();
plot2d(DAT(:,1),DAT(:,2),-3,rect=[-0.5,-0.5,2.5,2.5])        ← Fig. 4.2
t=linspace(-0.5,2.5,50);
et=p(1)*exp(p(2)*t)+p(3);
plot2d(t,et,rect=[-0.5,-0.5,2.5,2.5])
```

Note that we are using the same equations and the same initial guess as before. The value of p is completely different from the one we found before.

```
⊢⟶p
 p  =

! - 4154.231  - 0.0001505    4154.706 !
```

$\longmapsto$ v'
   ans   =

! - 0.4749930   0.5250070   - 0.1000375   - 0.2249880   0.2750120 !

However, the value of the norm squared is

$\longmapsto$ v'*v   ← norm squared
   ans   =

   0.6375094

which compares favorably with the 0.625 found by the earlier method. Looking at the graph and the data, we get Figure 4.2.

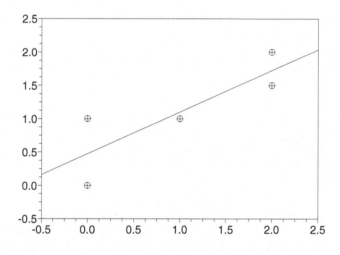

Figure 4.2. Data points and curve found by `lsqrsolve`.

The example of this section illustrates that problems can sometimes have several solutions, even when that is not obvious, and that different numerical methods, starting with the same starting value, can lead to very different solutions.

4.5 Parameter Fitting

We gave an example of data fitting in the previous section. But Scilab has some utilities that are especially designed for data fitting. We discuss here `datafit`, which is an improved version of the older routine `fit_dat`. For a given function $G(p, z)$, `datafit` finds the best vector of parameters p for approximating $G(p, z_i) = 0$ for a set of measurement vectors z_i. The vector p is found by minimizing $\sum_{i=1}^{n} G(p, z_i)'WG(p, z_i)$. That is, p is the solution of a weighted least squares problem. The most common use of W is to make some measurements more important than others. This utility uses `optim`, so the full calling sequence has many of the same optional variables.

The simplest call to `datafit` requires only the form of the equation G, the data, and a starting guess for the parameters p. In this case the calling sequence takes the form

$\longmapsto$ `[p,err]=datafit(G,Z,p0)`

Let `ne`, `np`, `nz` be the number of equations in `G`, the number of parameters in `p`, and the size of a single observation (dimension of `z`) respectively. Then

- `G` is a function describing `e=G(p,z)`.
- `Z` is the matrix of observations `[z_1,z_2,...,z_n]`, where each measurement is a column vector.
- `p0` is the initial guess for the parameters.

For example, if we wished to fit the data of the example in the last section, we could use the following Scilab script:

$\longmapsto$ `Z=[0 0 1 2 2;0 1 1 1.5 2];`

$\longmapsto$ `function e=G(p,z)`
$\longmapsto$ `    e=z(2)-p(1)*exp(p(2)*z(1))-p(3)`
$\longmapsto$ `endfunction`

$\longmapsto$ `p0=[0 0 0]';`

Note that the initial `p0` had to be put in as a column vector. When this script is executed we get the output

$\longmapsto$ `[p,err]=datafit(G,Z,p0);`

which is the same as that given by `leastsq` in the last section.

We saw in the last section that there can be more than one solution to a parameter-fitting problem. It can be important to bound the parameters to avoid values that are not physically correct. Also, sometimes some data are more important than other data. In this case one wants to weight the data. Sometimes the default values of parameters do not enable the algorithm to converge. These and other considerations sometimes require access to more of the parameters inside the algorithm. The full calling sequence, which shares a number of variables with `optim`, is

```
[p,err]=datafit([imp,] G [,DG],Z [,W],[contr],p0,[algo],[df0,[mem]],
                [work],[stop],['in'])
```

These parameters are as follows:

- `imp`: Scalar argument used to set the trace mode. `imp=0` means that nothing (except errors) is reported. `imp=1` means initial and final reports, `imp=2` adds a report per iteration and `imp>2` adds reports on linear search. Note that most of these reports are written on the Scilab standard output.
- `DG`: Partial of `G` with respect to `p`. In the form `S=DG(p,z)`, so that `S` is `ne`×`np`.
- `W`: Weighting matrix of size `ne`×`ne`. Default is the identity.

- **"in"**: Reserved keyword for initialization of parameters used when G is given as a Fortran or C routine.
- **p**: Optimal solution found, given as a column vector.
- **err**: Least squares error.

All the remaining parameters are identical to those in the general call to optim.

4.6 Linear and Quadratic Programming

4.6.1 Linear Programs

The basic linear programing problem is to minimize $p'x$ subject to linear constraints. These linear constraints can be equality or inequality constraints. The inequality constraints can be one-sided or bound constraints. The Scilab function to solve these problems is linpro. The general form of the problem is

$$\min_x p'x, \tag{4.12a}$$
$$C_1 x \leq b_1, \tag{4.12b}$$
$$c_i \leq x \leq c_s, \tag{4.12c}$$
$$C_2 x = b_2. \tag{4.12d}$$

The form of the calling sequence depends on how many of these types of constraints are present. The calling sequences are

```
[x,lagr,f]=linpro(p,C,b [,x0])
[x,lagr,f]=linpro(p,C,b,ci,cs [,x0])
[x,lagr,f]=linpro(p,C,b,ci,cs,me [,x0])
[x,lagr,f]=linpro(p,C,b,ci,cs,me,x0 [,imp])
```

The first three of these correspond to having the constraints (4.12b), (4.12b) and (4.12c), (4.12b)–(4.12d) respectively. Note that if there are no constraints at all, then the problem has no finite solution if p is not the zero vector.

These parameters have the following meaning:

- **p**: Real column vector of dimension n that contains the coefficients of the linear cost function.
- **C**: Real (me + md)×n matrix. If no constraints are given, that is, all constraints are of the form (4.12c), you can set C = []. Equality constraints are listed first, so that if both (4.12b) and (4.12d) are present, we have C(j,:)x = b(j) for j=1,...,me and C(j,:)x <= b(j), for j=me+1,...,me+md.
- **b**: (me + md)-dimensional column vector that is the right hand of the constraints in (4.12b), (4.12d). If no constraints are given, then you can set b = [].
- **ci**: n-dimensional column vector of lower bounds. If there are no lower bound constraints, put ci = []. If some components of x are bounded from below, set the other (unconstrained) values of ci to a negative number that has very large magnitude. That is, ci(j)=-number_properties('huge').
- **cs**: Column vector of upper bounds. See comments for ci.
- **me**: Number of equality constraints. Thus C(1:me,:)*x = b(1:me).
- **x0**: Either an initial guess for x or one of the character strings 'v' or 'g'. If x0='v', then the calculated initial feasible point is a vertex. If x0='g', then the calculated initial feasible point need not be a vertex.

- **imp**: This is the verbose option. (Try $imp = 7, 8 \dots$). Note that the messages are sent to Scilab standard output (can be lengthy if there are many iterations).
- **x**: Optimal solution found by algorithm.
- **f**: Optimal value of the cost function (4.12a).
- **lagr**: This is a vector of Lagrange multipliers. If lower and upper-bounds ci,cs are provided, lagr has n+me+md components and lagr(1:n) is the Lagrange vector associated with the bound constraints and lagr(n+1: n+me+md) is the Lagrange vector associated with the linear constraints. If an upper bound (respectively lower bound) constraint i is active, then lagr(i) is greater than 0 (resp. <0). If no bounds are provided, then lagr has only me+md components.

4.6.2 Quadratic Programs

A quadratic program is the same as (4.12) except that the cost function (4.12a) is replaced by the quadratic expression

$$\frac{1}{2}x'Qx + p'x. \tag{4.13}$$

The utility provided by Scilab 5.1 and ScicosLab 4.3 for solving this minimization problem is quapro. ScicosLab 4.4 will instead use qpsolve. Since qpsolve does not require Q to be positive definite, in fact Q could be zero, qpsolve can solve linear programming problems. With the exception of the matrix Q, which must be real symmetric, the calling sequence for qpsolve is identical to that of linpro. Since the variables have the same meanings, we do no repeat their definitions here.

In ScicosLab 4.4 quapro will be available in a toolbox. Users of quapro should note that if x0 is a scalar, and the full syntax is not used, then an error can result. That is because the possible calling sequences for quapro are

```
[x,lagr,f]=quapro(Q,p,C,b [,x0])
[x,lagr,f]=quapro(Q,p,C,b,ci,cs [,x0])
[x,lagr,f]=quapro(Q,p,C,b,ci,cs,me [,x0])
[x,lagr,f]=quapro(Q,p,C,b,ci,cs,me,x0 [,imp])
```

and quapro decides which sequence is being used by testing the last entry to see if it is a scalar. Thus if xo is a scalar, it can think it is me.

4.6.3 Semidefinite Programs

There is also a Scilab function for solving semidefinite programs called semidef:

```
[x,Z,ul,info]=semidef(x0,Z0,F,blck_szs,c,options)
```

However, we shall not discuss it here. The interested reader is referred to the on-line help.

4.7 Differentiation Utilities

The function NDcost was introduced earlier to enable the user to solve some problems without having to specify the gradient. Other utilities described earlier will automatically estimate derivatives or Jacobians numerically when this is requested. However, it is sometimes useful to be able to get numerical estimates of derivatives and Jacobians directly. This capability is provided by numdiff and derivative.

The Scilab function numdiff computes a numerical estimate of the Jacobian using finite difference methods. If the function has real-number values, then it gives an estimate of the gradient. Its calling sequence is

```
g=numdiff(fun,x [,dx])
```

if there are no parameters in `fun`. The parameters in `numdiff` are

- `fun`: An external describing the function to be differentiated. The function `fun` calling sequence must be `y=fun(x,p1,p2,...,pn)`, where the `pi` are optional parameters. If parameters `p1,p2,...,pn` exist, then `numdiff` can be called by `g=numdiff(list(fun,p1,p2,...,pn),x)`.
- `x`: Vector argument of the function `fun`.
- `dx`: Vector, which is the finite difference step. Default value is

    `dx=sqrt(%eps)*(1+1d-3*abs(x))`

- `g`: Vector (matrix), which is the estimated gradient (Jacobian).

For example, suppose that we want the Jacobian (derivative) of

$$f(x_1, x_2, x_3) = \begin{pmatrix} x_1 + 2x_2^2 x_3 \\ \sin(x_1 x_2 x_3) \end{pmatrix} \quad \text{at } x = \begin{pmatrix} 1 \\ 2 \\ 3 \end{pmatrix}. \tag{4.14}$$

The following Scilab script, when executed, computes the estimate of the Jacobian given by `numdiff`, the true Jacobian, and their difference.

```
function z=myf(x)
  z=[x(1)+2*x(2)-x(2)^2*x(3);sin(x(1)*x(2)*x(3))]
endfunction

x=[1;2;3];

J=numdiff(myf,x);
TrueJ=[1,2-2*x(2)*x(3), -x(2)^2];
a=cos(x(1)*x(2)*x(3));
TrueJ=[TrueJ;a*[x(3)*x(2),x(1)*x(3),x(1)*x(2)]];
Difference=J-TrueJ;
```

The solution is

```
⟼ J
 J  =

!   1.0000000  - 10.      - 4.        !
!   5.7610218    2.8805109   1.9203406 !

⟼ TrueJ
 TrueJ  =

!   1.        - 10.      - 4.        !
!   5.7610217    2.8805109   1.9203406 !

⟼ Difference
 Difference  =

! - 1.286D-08  - 4.045D-08    3.518D-08 !
!   1.157D-07    5.690D-08  - 1.100D-08 !
```

4.7.1 Higher Derivatives

Sometimes it is desirable to have higher derivatives. Numerical differentiation is prone to ill conditioning and error and this is especially true with the higher derivatives. If high-accuracy higher derivatives are required, the best alternatives are symbolic differentiation using a symbolic language such as Maple or automatic differentiation using a package such as ADOL-C. [28] For complicated problems the symbolic approach can be very slow and have memory problems. The only choice then may be automatic differentiation.

If the needed derivatives are limited to first and second order, then useful approximate derivatives can often by gotten by using the Scilab function `derivative`.

Suppose that we have a function such as (4.15) of an n-dimensional vector x with values that are m-dimensional vectors. Then at a point a it has a Taylor expansion

$$f(x) = \begin{pmatrix} f_1(x) \\ \vdots \\ f_m(x) \end{pmatrix} = f(a) + J(a)(x-a) + \begin{pmatrix} (x-a)'H_1(a)(x-a) \\ \vdots \\ (x-a)'H_m(a)(x-a) \end{pmatrix} + \cdots . \quad (4.15)$$

Thus the first derivative is just the $m \times n$ Jacobian matrix J. However, the second derivative is now m matrices each of which is $n \times n$. The first and second derivatives of f are provided by `derivative`. This utility provides several choices on how to arrange the information on the second derivative by use of the `H_form` variable. The full calling syntax is

`[J [,H]] = derivative(F,x [,h ,order ,H_form ,Q])`

- `F`: Scilab function `F` or a `list(F,p1,...,pk)`, where `F` is a Scilab function in the form `y=F(x,p1,...,pk)`, with `p1, ..., pk` being any Scilab objects such as matrices or lists.
- `x`: real column vector of dimension `n`.
- `h`: Step size used in the finite difference approximations. Usually better to leave to software to choose.
- `order`: Integer that is the order of the finite difference formula used to approximate the derivatives. Default is `order= 2`. May be chosen as $1, 2$, or 4.
- `H_form`: String giving the form in which the Hessian will be returned.
  - `H_form='default'`: H is an $m \times (n\verb|^|2)$ matrix. Here the kth row of H corresponds to the Hessian of the kth component of `F`.
  - `H_form='blockmat'`: In this format H is a $(m \times n) \times n$ block matrix . Thus, in Scilab notation, H = [H1;H2; ... ;Hm].
  - `H_form='hypermat'`: H is a $n \times n$ matrix for `m=1`, and a $n \times n \times m$ hypermatrix otherwise. That is, `H(:,:,k)` is the classical Hessian matrix of the kth component of `F`.
- `Q`: Real orthogonal matrix. Default is `eye(n,n)`.

The Jacobian is computed by approximating the directional derivatives of the components of `F` in the direction of the columns of `Q`. The second derivatives are computed by composition of first-order derivatives. Numerical approximation of derivatives is generally an unstable process. The step size `h` must be small to get a low error, but if it is too small, floating-point errors will dominate by cancellation. A rule of thumb is not to change the default step size. To work around numerical difficulties one may also change the order and/or choose different orthogonal matrices `Q` (the default is `eye(n,n)`), especially if the approximate derivatives are used in optimization routines. All the optional arguments may also be passed as named arguments.

As an illustration suppose that `myf` is the same function as in the previous example and that it and `x` are already in the workspace. Then the second derivative will have two 3×3 matrices. If we choose `'hypermat'` for the output format, then we will obtain

```
⟼   function z=myf(x)
⟼     z=[x(1)+2*x(2)-x(2)^2*x(3);sin(x(1)*x(2)*x(3))]
⟼   endfunction
⟼ x=[1;2;3];
⟼ [J,H]=derivative(myf,x,H_form='hypermat')
 H  =


(:,:,1)

!   0.    0.    0. !
!   0.  - 6.  - 4. !
!   0.  - 4.    0. !
(:,:,2)

!   10.058956    7.9099883    5.2733256 !
!   7.9099883    2.5147394    2.6366631 !
!   5.2733256    2.6366631    1.117662  !
 J  =

!   1.         - 10.        - 4.         !
!   5.7610217    2.8805109    1.9203406 !
```

5

Examples

Up to this point we have deliberately kept most of our examples concise in order to focus on learning Scilab and also to illustrate how simple many types of calculations are with Scilab. However, real scientific inquiries are often more complicated, involve several kinds of software, and have several goals.

In this chapter we provide a few more elaborate application examples to illustrate how Scilab can be used within this more complicated environment of investigation.

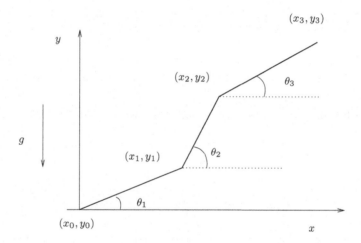

Figure 5.1. 3-link pendulum.

5.1 Modeling and Simulation of an N-Link Pendulum

Our first application will be the modeling and simulation of a pendulum with N links. With this problem we shall illustrate how a symbolic manipulation package, in this case Maple, can be used to derive the equations of motion of a multibody system and simulate the result in Scilab. This is done using the Lagrangian approach.

The N-link pendulum is a well-known problem in numerical analysis. For large N it is used as a test problem for software used for molecular dynamics simulation. In this case it is very important to know whether energy is conserved.

The N-link pendulum also arises as a model for flexible cables and chains. Several years ago there was a collapse of several towers carrying power. It turned out that the power lines were supported by pairs of chains of insulators and one of these chains had failed and the other was unable to continue to support the power line. In the followup studies, the insulator chains were modeled as 16-link pendulums.

5.1.1 Equations of Motion of the N-Link Pendulum

In this application we wish to simulate the motion of an N-link pendulum fixed at one extremity and moving without friction in a vertical plane. As an example, and for fixing notation, a picture of a 3-link pendulum is given in Figure 5.1. The equation of motion can easily be deduced from a Lagrangian formulation for a 1- or 2-link pendulum but it can become tedious work for say a 10-link pendulum in the Scilab demos.

In order to construct models with large N, we use the computer algebra package Maplein order to derive the equations of motion from a Lagrangian formulation. The computer algebra package generates Fortran code to be used by Scilab for simulation. It also provides LaTeX code for symbolic expressions at every step of the computation. The N-link pendulum is considered as N independant connected rigid bodies.

The Maple code is separated into two parts. The first part provides a general library for Lagrangian modeling of rigid bodies linked with holonomic constraints, and the second part applies the general code to the specific example of the N-link pendulum. By setting the code up in this manner, we have made it easier for the reader to modify the pendulum problem, say by adding masses at the end or even modeling a completely different mechanical system.

It should be noted that the kind of constraints we have here are called holonomic. Constraints involving velocities are called nonholonomic and are not considered here.

Lagrangian of Multibody Systems with Holonomic Constraints

Let q be the generalized coordinates and $\dot{q}$ the generalized velocities of the system of N rigid bodies. The coordinates q can vary with time t. Then the Lagrangian of the problem can be expressed as

$$\mathcal{L}(q,\dot{q}) = \sum_{i=1}^{N} \mathcal{L}_i(q,\dot{q}), \tag{5.1}$$

where $\mathcal{L}_i$ is the Lagrangian of body i. For each component of q we have to compute $\mathcal{I}(q,\dot{q},\ddot{q})_i$, which is defined by the Euler equation

$$\mathcal{I}(q,\dot{q},\ddot{q})_i = \frac{d}{dt}\left(\frac{\partial \mathcal{L}}{\partial \dot{q}^i}\right) - \frac{\partial \mathcal{L}}{\partial q^i}.$$

Readers familiar with the calculus of variations or optimal control will recognize this as the usual term that arises in the necessary conditions for a minimum.

This formal expression for $\mathcal{I}(q)$ can be rewritten in matrix notation as

$$\mathcal{I}(q,\dot{q},\ddot{q}) = M(q)\ddot{q} + R(q,\dot{q}). \tag{5.2}$$

If there are no constraints or outside forces, then (5.2) is set equal to zero, and we get the usual second-order equation of motion, which says that mass times acceleration is equal to the total force.

So far we have assumed that the rigid bodies were independent. We now have to deal with the mechanical constraints of holonomic type expressing the fact that we have a linked pendulum. These constraints can be written as $A(q) = 0$. Note that the time derivative of the constraints gives the equation

$$B(q)\dot{q} = 0,$$

where $B(q) = \frac{dB}{dq}$, that is,

$$B(q)_{i,j} = \left(\frac{\partial A(q)_i}{\partial q_j} \right).$$

Thus $\dot{q}$ has to be in the kernel of $B(q)$. We can then try to compute a formal basis of the kernel of $B(q)$. This yields a matrix $S(q)$ whose columns span the kernel of $B(q)$. Then we can introduce a time function η such that $\dot{q} = S(q)\eta$. Note that by definition we have

$$B(q)S(q)\eta = 0. \tag{5.3}$$

We return now to the Euler equations. Using the constraints and their associated Lagrange multipliers, the equations can be expressed as

$$\mathcal{I}(q, \dot{q}, \ddot{q}) = B(q)^T \lambda. \tag{5.4}$$

We use $S(q)$ to eliminate the right-hand side of (5.4) by left multiplying by $S(q)^T$ and simplifying the result using $\dot{q} = S(q)\eta$ and its time derivative, which can be expressed as $\ddot{q} = W(q, \eta, \dot{\eta})$. We finally obtain

$$0 = S(q)^T \mathcal{I}(q, \dot{q}, \ddot{q}) = M_1(q)\dot{\eta} + R_1(q, \eta).$$

The dynamical system with state (q, η) to be simulated is then

$$M_1(q)\dot{\eta} + R_1(q, \eta) = 0, \tag{5.5a}$$
$$\dot{q} - S(q)\eta = 0. \tag{5.5b}$$

Note that in this last equation the constraint $A(q) = 0$ is seen only through its time derivative. This raises the possibility that during simulation there may be numerical errors in the constraints. In the numerical analysis literature this is called constraint drift. However, we will see that in the example of the N-link pendulum, we can simplify the equations in order to avoid the numerical drift in the constraint $A(q) = 0$.

When it is not possible to find a formal basis of the null space of $B(q)$, it is often possible to differentiate the holonomic constraint twice, which leads to the constraint

$$B(q)\ddot{q} + B_q(q)\dot{q} = 0.$$

Thus we have $A(q) = 0$, $B(q)\dot{q} = 0$, and $B(q)\ddot{q} + B_q(q)\dot{q} = 0$. These three constraints can be replaced by a unique constraint [9]

$$-B(q)\ddot{q} + R_2(q, \dot{q}) = 0$$

with

$$R_2(q, \dot{q}) = -B_q(q)\dot{q} + \alpha \left(B(q)\dot{q} \right) + \beta A(q) = 0.$$

This replacement process is known as stabilized differentiation or Baumgarte stabilization. The new system to be simulated is then

$$M(q)\ddot{q} - B(q)^T \lambda + R(q, \dot{q}) = 0, \tag{5.6a}$$
$$-B(q)\ddot{q} + R_2(q, \dot{q}) = 0. \tag{5.6b}$$

At each integration step we need to solve a linear system to get $(\ddot{q}, \lambda)$, and α and β are to be properly chosen so as to keep the error stabilized. A similar problem has been considered in [24] where the control of a bicycle was considered.

Note that neither (5.5) nor (5.6) is an explicit ordinary differential equation. Depending on the invertibility of M_1 and B they are either implicit ordinary differential equations or differential algebraic equations (DAE). There are two options. One is to try to integrate with a DAE solver such as `dassl`. Alternatively, one can try to reduce the system to an ordinary differential equation. We shall do the latter.

The N-Link Pendulum as N Rigid Bodies with Holonomic Constraints

We consider here an N-link pendulum that is fixed at one end and moves without friction in a vertical plan as described in Figure 5.1. In this section we derive, with specialized Maple code, the equations of motion of the N-link pendulum. We will follow the steps of the previous section. In order to keep the output simple, we will use the simpler 2-link pendulum. But the reader should keep in mind that the equations are not manually written but rather are derived by a computer program.

We first generate the Lagrangian of the N-link pendulum by adding the Lagrangian of each rigid body. Let $r_i = l_i/2$. For the 2-link pendulum this gives

$$
\begin{aligned}
\mathcal{L}(q,\dot{q}) = \frac{1}{2}m_1 \left(r_1{}^2 \sin(\theta_1)^2 \dot{\theta}_1^2 + r_1{}^2 \cos(\theta_1)^2 \dot{\theta}_1^2 \right) + \frac{J_1 \dot{\theta}_1^2}{2} - m_1 \, g \, r_1 \, \sin(\theta_1) \\
+ \frac{1}{2}m_2 \left(\left(\dot{x}_1 - r_2 \sin(\theta_2) \dot{\theta}_2 \right)^2 + \left(\dot{y}_1 + r_2 \cos(\theta_2) \dot{\theta}_2 \right)^2 \right) + \frac{J_2 \dot{\theta}_2^2}{2} \\
- m_2 \, g \, (y_1 + r_2 \sin(\theta_2)).
\end{aligned}
$$

Then we obtain the following expression for $\mathcal{I}(q, \dot{q}, \ddot{q})$:

$$
\begin{pmatrix}
m_2 \ddot{x}_1 - m_2 r_2 \cos(\theta_2) \dot{\theta}_2^2 - m_2 r_2 \sin(\theta_2) \ddot{\theta}_2 \\
m_2 \ddot{y}_1 - m_2 r_2 \sin(\theta_2) \dot{\theta}_2^2 + m_2 r_2 \cos(\theta_2) \ddot{\theta}_2 + m_2 g \\
m_1 r_1{}^2 \ddot{\theta}_1 + J_1 \ddot{\theta}_1 + m_1 g r_1 \cos(\theta_1) \\
0 \\
0 \\
-m_2 r_2 \sin(\theta_2) \ddot{x}_1 + m_2 r_2{}^2 \ddot{\theta}_2 + m_2 r_2 \cos(\theta_2) \ddot{y}_1 + J_2 \ddot{\theta}_2 + m_2 g r_2 \cos(\theta_2)
\end{pmatrix},
$$

which is rewritten as in (5.2) with

$$
M(q) = \begin{pmatrix}
m_2 & 0 & 0 & 0 & 0 & -\alpha_2 \\
0 & m_2 & 0 & 0 & 0 & \beta_2 \\
0 & 0 & m_1 r_1{}^2 + J_1 & 0 & 0 & 0 \\
0 & 0 & 0 & 0 & 0 & 0 \\
0 & 0 & 0 & 0 & 0 & 0 \\
-\alpha_2 & \beta_2 & 0 & 0 & 0 & m_2 r_2{}^2 + J_2
\end{pmatrix},
$$

$$
R(q, \dot{q}) = \begin{pmatrix}
-m_2 r_2 \cos(\theta_2)\dot{\theta}_2{}^2 \\
m_2 g - m_2 r_2 \sin(\theta_2)\dot{\theta}_2{}^2 \\
m_1 g r_1 \cos(\theta_1) \\
0 \\
0 \\
m_2 g r_2 \cos(\theta_2)
\end{pmatrix},
$$

where $\beta_k = m_k r_k \cos(\theta_k)$ and $\alpha_k = m_k r_k \sin(\theta_k)$.

We have simplified the system by eliminating (x_0, y_0) and $(\dot{x}_0, \dot{y}_0)$, since one end of the first link is at a fixed position.

The constraints are

$$x_1 - 2r_1 \cos(\theta_1) = 0, \tag{5.7a}$$
$$y_1 - 2r_1 \sin(\theta_1) = 0, \tag{5.7b}$$
$$x_2 - x_1 - 2r_2 \cos(\theta_2) = 0, \tag{5.7c}$$
$$y_2 - y_1 - 2r_2 \sin(\theta_2) = 0. \tag{5.7d}$$

In general we would have $2N$ constraints for a nN-link pendulum, which are given by $A(q) = 0$.

Symbolic differentiation of $A(q)$ by Maple leads to $B(q)$:

$$B(q) = \begin{pmatrix} 1 & 0 & 2r_1 \sin(\theta_1) & 0 & 0 & 0 \\ 0 & 1 & -2r_1 \cos(\theta_1) & 0 & 0 & 0 \\ -1 & 0 & 0 & 1 & 0 & 2r_2 \sin(\theta_2) \\ 0 & -1 & 0 & 0 & 1 & -2r_2 \cos(\theta_2) \end{pmatrix}.$$

A symbolic computation for a basis of the kernel of $B(q)$ gives

$$S(q) = \begin{pmatrix} -2\,r_1 \sin(\theta_1) & 0 \\ 2\,r_1 \cos(\theta_1) & 0 \\ 1 & 0 \\ -2\,r_1 \sin(\theta_1) & -2\,r_2 \sin(\theta_2) \\ 2\,r_1 \cos(\theta_1) & 2\,r_2 \cos(\theta_2) \\ 0 & 1 \end{pmatrix}.$$

As pointed out in the previous section we can thus introduce a time function η such that $\dot{q} = S(q)\eta$. Using the particular form of $S(q)$, we note that the η variable has a physical interpretation as $\eta = \dot{\theta}$. Note also that $S(q)$ is not unique and we have selected the $S(q)$ for which we have

$$\eta = \dot{\theta}.$$

We use $S(q)$ to eliminate the right-hand side of this equation by left multiplying by $S(q)^T$ and simplifying the result using $\dot{q} = S(q)\eta$. Using basic symbolic algebra, the obtained equation can be rewritten as

$$M_1(q)\dot{\eta} + R_1(q, \eta) = 0,$$

where

$$M_1(q) = \begin{pmatrix} m_1\,r_1{}^2 + J_1 + 4\,r_1{}^2\,m_2 & 2\,r_1\,m_2\,r_2\,\cos(\theta_1 - \theta_2) \\ 2\,r_1\,m_2\,r_2\,\cos(\theta_1 - \theta_2) & m_2\,r_2{}^2 + J_2 \end{pmatrix},$$

$$R_1(q, \eta) = \begin{pmatrix} r_1\,\cos(\theta_1)\,g\,(2\,m_2 + m_1) + 2\,r_1\,m_2\,r_2\,\sin(\theta_1 - \theta_2)\eta_2^2 \\ m_2\,g\,r_2\,\cos(\theta_2) - 2\,r_1\,m_2\,r_2\,\sin(\theta_1 - \theta_2)\eta_1^2 \end{pmatrix}.$$

The dynamical system with state (q, η) to be simulated is then

$$M_1(q)\dot{\eta} + R_1(q, \eta) = 0,$$
$$\dot{q} - S(q)\eta = 0.$$

Note that M_1, R_1, and S depend only on θ and η. Thus we can drop $(x_i, y_i)_{i=1,2}$ from the system equations and just keep (θ, η) as the state, giving us

$$M_1(\theta)\dot{\eta} + R_1(\theta, \eta) = 0, \tag{5.8a}$$
$$\dot{\theta} = \eta. \tag{5.8b}$$

Notice that (5.8) has the expected form with θ being the angular position and η being the angular velocity.

When a solution is found for (θ, η), it is easy to recompute the $(x_i, y_i)_{i=1,2}$ positions using the constraints (5.7). This calculation works in general and we have shown that the dynamics of the N-link depend only on the state variables $(\theta, \dot{\theta})$.

5.1.2 Generated Code and Simulation

Using the Maple computer algebra system we can code as a Maple script the steps described in the previous section. This script was used to generate the three Fortran files: npend.f (dynamics), np.f (number of links), and ener.f (total energy), and LaTeX code of the equations that were placed, using cut and paste, in the previous paragraph.

We provide in listing (5.1.2) a verbatim input of the Fortran file npend.f containing npend(neq,t,th,ydot), which codes the dynamics given by (5.1.2). The equations given by (5.1.2) are in implicit form, and we numerically solve a linear system in npend ($M_1(q)\dot{\eta} = -R_1(q, \eta)$) in order to compute $\dot{\eta}$. Note that this file is 59 lines long for a 2-link pendulum. For a 10-link pendulum the generated file, which can be found in the Scilab distribution at demos/simulation/npend/Maple, is 477 lines long!

```
c    SUBROUTINE npend
     subroutine npend(neq,t,th,ydot)
     parameter (n=2)
     implicit doubleprecision (t)
     doubleprecision t,th(2*n),eta(n),ydot(2*n),r(n),
     doubleprecision me3s(n,n),const(n,1),j(n),m(n)
     doubleprecision w(3*n),rcond
     integer i,k,neq,ierr
     data g / 9.81/
     data r / n*1.0/
     data m / n*1.0/
     data j / n*0.3/
c
     do 1000, i =1,n ,1
        ydot(i) = th(i+n)
1000 continue
c
c
     do 1001, i =1,n ,1
        eta(i) = th(i+n)
1001 continue
c
     t1 = r(1)**2
     t8 = cos(th(1)-th(2))
     t11 = 2*r(1)*m(2)*r(2)*t8
     t12 = r(2)**2
     me3s(1,1) = m(1)*t1+4*t1*m(2)+J(1)
     me3s(1,2) = t11
     me3s(2,1) = t11
     me3s(2,2) = m(2)*t12+J(2)
```

```
      t1 = m(2)*r(2)
      t2 = eta(2)**2
      t4 = sin(th(1)-th(2))
      t8 = cos(th(1))
      t16 = eta(1)**2
      t20 = cos(th(2))
      const(1,1) = r(1)*(2*t1*t2*t4+2*t8*m(2)*g+m(1)*g*t8)
      const(2,1) = t1*(-2*r(1)*t16*t4+g*t20)
c
      do 1002, i =1,n ,1
         const(i,1) = -const(i,1)
 1002 continue
c
c     solving M z = const to obtain ydot((n+1)..2*n)
      call dlslv(me3s,n,n,const,n,1,w, rcond,ierr,1)
      if (ierr.ne.0) then
         write(6,2000)
 2000    format('Matrix is badly conditioned')
      endif
c
      do 1003, i =1,n ,1
         ydot(n+i) = const(i,1)
 1003 continue
c
      return
      end
```

As described in Section 2.5.1 on interfacing, it is now possible to link the function
`npend` (which conforms to the internal calling syntax accepted by `ode`), and numerically
integrate the *N*-pendulum equation using `ode`. The `np` function, which is also dynamically
linked in Scilab, returns the number of links of the pendulum. The reader can take a look
at the `SCI/demos/simulation/npend` to check the details. The Fortran code is compiled
and dynamically linked in a running Scilab using the previously described `ilib_for_link`
function.

```
⟼ npend_build_and_load() ;                          ← calling ilib_for_link
   generate a loader file
   generate a Makefile: Makelib
   running the makefile
   compilation of npend
   compilation of np
   compilation of ener
   compilation of dlslv
   building shared library (be patient)
shared archive loaded
Link done

⟼ n=np(); r=ones(1,n); m=ones(1,n); j=ones(1,n); g=9.81;
⟼ y0=0*ones(2*n,1);tt=0:0.05:10;                     ← (θ,θ̇)(0) = 0
⟼ yt=ode(y0,0,tt,'npend');                           ← calling ode
⟼ draw_chain_from_angles(yt(1:$/2,:),r,1);           ← Figure~5.2
```

Then the N-link pendulum is easily simulated using `ode` given initial values for $(\theta, \dot{\theta})$. Using the scilab demo files, we now show the result (See Figure 5.2) of a simulation of a 10-link pendulum.

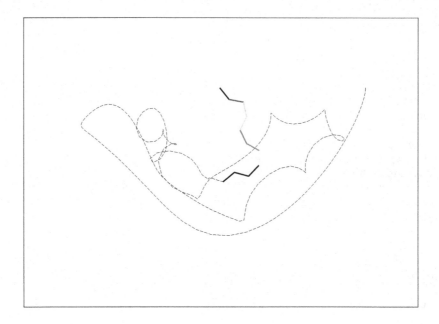

Figure 5.2. 10-link pendulum. Dotted line represents the trajectory of the endpoint of the pendulum.

The result of the `ode` simulation is a matrix with N rows. Each column of the matrix gives the value of $(\theta, \dot{\theta})$ at a specific time. Using this matrix a graphics simulation is obtained by first computing the $(x_i, y_i)_{i=0,N}$ and then using the function `draw_chain_from_angles` to provide an animation. There is an option in the animation to keep track of the trajectory of the endpoint of the pendulum. This curve can be seen as the dashed line in the static Figure 5.2.

We provide here a slightly modified version of `draw_chain_from_angles`, which gives an example of an animation using the new graphics mode. Moreover, this function uses the Scilab functions `realtimeinit` and `realtime` to control the animation speed.

```
function draw_chain_from_angles(a,r,job,rt)
// a the angles , a(i,j) is the angle of node i at time t(j)
// r the segments half length
   if argn(2)<3  then job=0, end
   if argn(2)<4  then rt = 0.01; end
   n2=size(a,2);
   x=[0*ones(1,n2);cumsum(2*diag(r)*cos(a),1)];
   y=[0*ones(1,n2);cumsum(2*diag(r)*sin(a),1)];
   draw_chain_from_coordinates(x,y,job,rt)
endfunction

function draw_chain_from_coordinates(x,y,job,rt)
// x(i,j), y(i,j) is the coordinate of node i at time t(j)
```

```
// r the segments half length
  if argn(2)<3  then job=0, end
  if argn(2)<4  then rt = 0.01; end
  [n1,n2]=size(x);
  b = maxi(maxi(abs(x)),maxi(abs(y)));
  xbasc();set figure_style new;                          ← new graphics mode
  a=gca()                                                ← we want to set
  a.data_bounds=[-1,-1;1,0]*b;                       ← the graphics boundaries
  a.isoview="on";                                    ← and use an isoview mode
  xset('pixmap',1);                                     ← double buffer mode
  drawlater()                                        ← wait for graphics display
  colors= 1:n1-1; colors(8)=n1;
  xsegs([x(1:$-1,1)';x(2:$,1)'],[y(1:$-1,1)';y(2:$,1)'],colors)          ← draw
  p=gce();p.thickness=4;                       ← p is used to keep track of xsegs data
  if job==1  then
     xpoly(x($,1)*ones(2,1),y($,1)*ones(2,1),'lines');       ← endpoint trajectory
     t=gce();t.line_style=2;
  end
  drawnow()                                              ← perform drawing
  xset('wshow')                              ← update display with double buffer
  ind=[1;(2:n1-1)'.*.ones(2,1);n1]
  realtimeinit(rt);           ← set time unit: smaller values will accelerate the animation
  for j=1:n2,
    realtime(j)                              ← wait to time j before continuing
    drawlater()
    p.data = [x(ind,j),y(ind,j)];                          ← update xsegs data
    if job==1  then t.data=[t.data;[x($,j),y($,j)]], end     ← update xpoly data
    drawnow()                     ← since objects are updated,  graphics is cleared and redrawn
    xset('wshow')                            ← update display with double buffer
  end
endfunction
```

5.1.3 Maple Code

General Code Dealing with Euler Equation

We provide here part of the file `Euler.map`, which contains generic functions for Euler equation computation from Lagrangian formulation. For simplicity we have removed from the Maple code functions devoted to LaTeX code generation, but the full file content can be found in the Scilab distribution.

```
#-------------------------------------------------------------------
# Functions for computing Euler equations.  L is the Lagrangian,
# q,qd,qdd are the generalized coordinates and their derivatives
#-------------------------------------------------------------------

with('linalg'):

euler_equations:=proc(L,q,qd,qdd)
       local k,m:
       m:=nops(q);
       v:=matrix(m,1,0);
       for i to m  do
```

```
            v[i,1]:=LL(q[i])=simplify(time_diff(diff(L,qd[i]),q,qd,qdd)
                              -diff(L,q[i]));
        od;
        eval(v):
        end:


#----------------------------------------------------------------
# Time derivative computation of an expression
# depending on q,qd,qdd.  Used to compute Euler equations
#----------------------------------------------------------------

ttvar:=proc(xx)
        if type(xx,'indexed')
        then cat(op(0,xx),'d')[op(xx)]
        else cat(xx,'d') fi
end:

time_diff:=proc(phi,q,qd,qdd)
        local phi_copy,k,diff_phi:
        # substitution to specify that q,qd ,qdd depends on time
        phi_copy:=phi:
        phi_copy:=subs(map( xx-> xx=xx(t),[op(q),op(qd)]),phi_copy):
        diff_phi:=diff(phi_copy,t):
        # substitution to come back to our variables
        diff_phi:=subs(map(xx->diff(xx(t),t)=ttvar(xx),[op(q),op(qd)]),
                       diff_phi):
        diff_phi:=subs(map(xx->xx(t)=xx,[op(q),op(qd),op(qdd)]),diff_phi):
end:



#-------------------------------------------------------
# Rewriting the Euler equations to have a canonical form
#               ..          .
# El= ME(q)   q + RE(q,q)
# Computation of ME
# CEuler returns a list [ME,RE];
#-------------------------------------------------------

CEuler:=proc(E,q,qd,qdd)
        local Me,Ce,Re:
        Me:=MME(E,q,qd,qdd):
        Re:=RRE(E,Me,q,qd,qdd):
        [eval(Me),eval(Re)]:
        end:

MME:=proc(E,q,qd,qdd)
        local E1:
        E1:=eval(E):
        genmatrix([seq(E1[i,1],i=1..nops(qdd))],qdd):
        end:

#-----------------------------------------------------
#                      .
# Extract the RE(q,q) matrix  El= ME(q) qdd + RE(q,qd)
```

```
#--------------------------------------------------------

RRE:=proc(E,ME,q,qd,qdd)
        local MM:
        MM:=matadd(     E,multiply(ME,matrix(nops(q),1,qdd)),1,-1);
        MM:=map((x)-> simplify(x),eval(MM)):
        end:


#--------------------------------------------------------
# FORTRAN GENERATION
#--------------------------------------------------------

Gener:=proc(filename,fortranlist)
        global optimized;
        init_genfor();
        optimized:=true:
        writeto(filename):
        genfor(flist):
        writeto(terminal):
        end:
```

5.2 Modeling and Simulation of a Car

This example is presented to illustrate the use of the zero-crossing option in the ode function. is also particularly interesting because it is hybrid in nature. This means that it has different behaviors in different "states." In this example the different states are characterized by whether or not the wheels of the car are on the ground. Since our 2D model has only two wheels, the system has four states, since each wheel can be on or off the ground.

The term "state" is being used two ways in this section. If we wanted to be more precise, we would say that the state of the system is composed of both discrete and continuous states. The discrete part takes on four values depending on the contact of the wheels. For each value of the discrete state we have continuous state variables that describe the car. In the model that follows there are the same number of continuous state variables for each value of the discrete state variable. That need not be true in general.

We consider a very simple planar mechanical model so that the equations do not become very complex. We will make several simplifying assumptions in order to reduce the complexity of the model, yet keep the essence of the application so that the model's dynamic behavior remains realistic. It would be easy to modify this example to make it even more realistic.

5.2.1 Basic Model

The simple 2D model of the car that we consider here includes (see Figure 5.3):

- two parallel suspensions which are massless but have internal damping,
- massless wheels of size zero rolling on the ground with no loss of energy due to friction,
- a mass m with rotational inertia J in the middle of the massless rod connecting the top of the suspensions. The length of the rod connecting the top of the suspensions is a constant denoted by l.

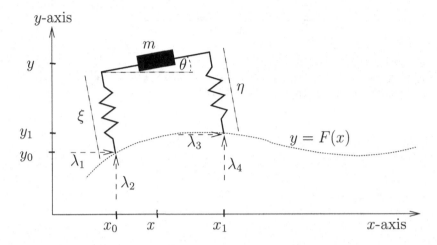

Figure 5.3. Simple model of the 2D car.

The model uses the following variables:

- θ: angle the car body makes with the horizonal. This is equal to the angles of the suspensions with the vertical.
- ξ and η: length of the two suspensions.
- λ_1 and λ_2 (respectively λ_3 and λ_4): the horizontal and vertical forces exerted by the ground on the rear (respectively front) wheel.
- ground modeled by the function $y = F(x)$.
- (x_0, y_0) (respectively (x_1, y_1)), which represent the coordinates of the rear (respectively front) wheel. They may or may not be on the ground (i.e., on the curve $F(x)$).
- (x, y): coordinate of the center of gravity of the mass m (the only object with mass in the model).

The two suspensions are assumed to have the parameters k, γ, and d. The forces produced by the suspensions are $k(\xi - d) + \gamma\dot{\xi}$ for the rear suspension and $k(\eta - d) + \gamma\dot{\xi}$ for the front suspension.

5.2.2 Equations of Motion

The equations of motion are obtained from basic principles of physics. We calculate the sum of external forces in the x and y directions to obtain the acceleration of the mass m in each of these directions. We also compute the torque around its center of gravity to obtain an equation for $\ddot{\theta}$. The remaining equations are obtained from the dynamics of the suspensions.

Since in our model each wheel can be on or above the ground, normally we should derive four different sets of equations describing the motion of the car. However, it turns out that once we have the model with both wheels on the ground, the other models can be obtained from it very easily. So we start with the assumption that both wheels are on the ground.

The equations of motion for the system are easily obtained from basic principles:

$$m\ddot{x} = \lambda_1 + \lambda_3, \tag{5.9a}$$

$$m\ddot{y} = -mg + \lambda_2 + \lambda_4, \tag{5.9b}$$

$$0 = k(\xi - d) + \gamma\dot{\xi} - \lambda_1\sin(\theta) + \lambda_2\cos(\theta), \tag{5.9c}$$

$$0 = k(\eta - d) + \gamma\dot{\eta} - \lambda_3\sin(\theta) + \lambda_4\cos(\theta), \tag{5.9d}$$

$$J\ddot{\theta} = \lambda_1\phi_0 - \lambda_2\psi_0 + \lambda_3\phi_1 + \lambda_4\psi_1, \tag{5.9e}$$

where

$$\phi_0 = \xi\cos(\theta) + l\cos(\theta)/2, \tag{5.9f}$$

$$\phi_1 = \eta\cos(\theta) - l\sin(\theta)/2, \tag{5.9g}$$

$$\psi_0 = -\xi\sin(\theta) + l\cos(\theta)/2, \tag{5.9h}$$

$$\psi_1 = \eta\sin(\theta) + l\cos(\theta)/2. \tag{5.9i}$$

System (5.9) gives a complete characterization of the dynamics of the system if the λ's are known.

In the absence of friction, the ground can apply only an orthogonal force to the wheels (the car does not accelerate or brake in our model). Thus

$$\lambda_1 = -\lambda_2 F_x(x_0), \tag{5.10}$$

$$\lambda_3 = -\lambda_4 F_x(x_1), \tag{5.11}$$

where

$$F_x(x) = \frac{\partial f}{\partial x}(x). \tag{5.12}$$

This means that the only values we need to determine are λ_2 and λ_4. Note that if a wheel is not in contact with the ground, then the corresponding λ's are simply zero.

If the wheels touch the ground, we have

$$y_0 = F(x_0), \tag{5.13}$$

$$y_1 = F(x_1). \tag{5.14}$$

The state of the car is completely characterized in terms of the variables x, y, θ, their derivatives, and ξ and η. Thus the wheel positions can be obtained from them:

$$x_0 = x - \psi_0, \tag{5.15}$$

$$y_0 = y - \phi_0, \tag{5.16}$$

$$x_1 = x + \psi_1, \tag{5.17}$$

$$y_1 = y - \phi_1. \tag{5.18}$$

The values of λ_2 and λ_4 can be obtained from (5.13)–(5.18) and the derivatives of (5.9c) and (5.9d):

$$\lambda_2 = \gamma(F_x(x_0)\dot{x} - w_0 k(\xi - d)/\gamma + (F_x(x_0)\phi_0 + \psi_0)\dot{\theta} - \ddot{y})/w_0^2, \tag{5.19}$$

$$\lambda_4 = \gamma(F_x(x_1)\dot{x} - w_1 k(\eta - d)/\gamma + (F_x(x_1)\phi_1 - \psi_1)\dot{\theta} - \ddot{y})/w_1^2, \tag{5.20}$$

where

$$w_i = F_x(x_i)\sin(\theta) + \cos(\theta), \quad i = 0, 1. \tag{5.21}$$

5.2.3 Simulation Model

To use the Scilab function `ode` to simulate this system, we need to put the system into a first order ODE. This is done by introducing variables v_x, v_y, and ω to represent the derivatives of x, y and θ. The first-order ODE can now be expressed as follows:

$$\dot{x} = v_x, \tag{5.22}$$
$$\dot{v}_x = -(\lambda_2 F_x(x_0) + \lambda_4 F_x(x_1))/m, \tag{5.23}$$
$$\dot{y} = v_y, \tag{5.24}$$
$$\dot{v}_y = -(mg - \lambda_2 - \lambda_4)/m, \tag{5.25}$$
$$\dot{\xi} = -(k(\xi - d) + \lambda_2 w_0)/\gamma, \tag{5.26}$$
$$\dot{\eta} = -(k(\eta - d) + \lambda_4 w_1)/\gamma, \tag{5.27}$$
$$\dot{\theta} = \omega, \tag{5.28}$$
$$\dot{\omega} = -(\lambda_2(F_x(x_0)\phi_0 + \psi_0) + \lambda_4(F_x(x_1)\phi_1 - \psi_1))/J, \tag{5.29}$$

where λ_2 and λ_4 are either zero (wheel not on the ground) or given by

$$\lambda_2 = \gamma(F_x(x_0)v_x - w_0 k(\xi - d)/\gamma + (F_x(x_0)\phi_0 + \psi_0)\omega - v_y)/w_0^2, \tag{5.30}$$
$$\lambda_4 = \gamma(F_x(x_1)v_x - w_1 k(\eta - d)/\gamma + (F_x(x_1)\phi_1 - \psi_1)\omega - v_y)/w_1^2. \tag{5.31}$$

The w's, ϕ's, and ψ's are defined as before.

Discrete States

Note that the system can be in 4 different states:

1. both wheels on the ground,
2. only rear wheel on the ground,
3. only front wheel on the ground,
4. no wheel on the ground.

The simulation must detect when the state changes. This is done by zero-crossing tests (the `root` option in `ode`).

If the system is in state 1 (both wheels on the ground), then the zero-crossing surfaces are

$$s = \begin{pmatrix} \lambda_2 \\ \lambda_4 \end{pmatrix}. \tag{5.32}$$

The reason is that as long as the wheels are on the ground, the vertical forces exerted on them from the ground are positive. As soon as any of them goes negative, the corresponding wheel leaves the ground.

If the system is in state 2 (only rear wheel on the ground), then the test on the rear wheel (λ_2) remains the same, but the other zero-crossing consists in finding whether the front wheel has come to contact with the ground. So in this case the zero-crossing surfaces are

$$s = \begin{pmatrix} \lambda_2 \\ y_1 - F(x_1) \end{pmatrix}. \tag{5.33}$$

Similarly, for state 3 the zero-crossing surfaces are

$$s = \begin{pmatrix} y_0 - F(x_0) \\ \lambda_4 \end{pmatrix}, \tag{5.34}$$

and for state 4,

$$s = \begin{pmatrix} y_0 - F(x_0) \\ y_1 - F(x_1) \end{pmatrix}. \tag{5.35}$$

5.2.4 Scilab Implementation

Now that the first-order dynamics and the zero-crossing surfaces are explicitly defined, the implementation in Scilab is straightforward using the ode function. The four states of the system are coded in two variables ST0 and ST1, which can take the values 0 and 1. The value 0 for ST0 indicates rear-wheel contact with the ground, while the value 0 for ST1 indicates front-wheel contact. A value of one means that the wheel is not in contact with the road.

Since the root option in ode does not allow the system to start off a zero-crossing surface, the simulation starts (and restarts after each zero-crossing detection) with a call to ode without the root option for a short period of time. Subsequently, ode with root option is called to perform the simulation up to the next zero-crossing. This also eliminates multiple zero-crossing detections for the same surface due to numerical errors.

The Scilab function simul_car, given below, implements the main simulation routine. This function receives the initial state X and a vector of time instances TT. It computes the solution XX of the system over the instances T. T is just TT to which the zero crossing instances have been added in order to capture important positions of the car.

```
function [XX,T]=simul_car(TT,X)
  Tt=TT; // remaining time instances
  XX=[];T=[]; // intialization of outputs
  tc=0;  //current time
  delt=1d-3; // time interval without root
  while 1
    t=tc+delt;
    It=find(Tt<t); // find indices of output points
    if It==[] then
      xt=ode(X,tc,t,car);  // no root simulation
    else
      Xx=ode(X,tc,Tt(It),car);
      T=[T,Tt(It)];XX=[XX,Xx];xt=Xx(:,$);Tt(It)=[];
    end
    [Xx,rd]=ode('root',xt,t,Tt,car,2,carg);
    if rd==[] then // no root found, update outputs
      XX=[XX,Xx];T=[T,Tt]; break;
    end
    tc=rd(1);  // root found at time tc
    if size(rd,'*')==3 then // two roots crossed
      ST0=1-ST0;ST1=1-ST1; // change both states
    else
      if rd(2)==1 then ST0=1-ST0; end  // rear wheel
      if rd(2)==2 then ST1=1-ST1; end  // front wheel
    end
    T=[T,Tt(find(Tt<rd(1))),rd(1)];XX=[XX,Xx];X=Xx(:,$);
    // update outputs, time of zero-crossing added to TT
    Tt=Tt(find(Tt>rd(1))); // update remaining time instances
  end
endfunction
```

The car model `car` and the zero-crossing surfaces `carg` functions are defined as follows:

```
function Xd=car(t,X)
 x=X(1);xd=X(2);y=X(3);yd=X(4);th=X(5);thd=X(6);xsi=X(7);eta=X(8);
 [lam2,lam4,x0,x1,mu0,mu1,w0,w1]=fun(X);
 if ST0  then lam2=0; end
 if ST1  then lam4=0; end
 xdd=-(lam2*Fx(x0)+lam4*Fx(x1))/m
 ydd=-(m*g-lam2-lam4)/m
 thdd=-(lam2*mu0+lam4*mu1)/J
 xsid=-(k*(xsi-d)+lam2*w0)/gam
 etad=-(k*(eta-d)+lam4*w1)/gam
 Xd=[xd;xdd;yd;ydd;thd;thdd;xsid;etad]
endfunction

function s=carg(t,X)
 x=X(1);xd=X(2);y=X(3);yd=X(4);th=X(5);thd=X(6);xsi=X(7);eta=X(8);
 [lam2,lam4,x0,x1,mu0,mu1,w0,w1]=fun(X);
 if ST0  then
   s1=y-xsi*cos(th)-l2*sin(th)-F(x0)
 else
   s1=lam2
 end
 if ST1  then
   s2=y-eta*cos(th)+l2*sin(th)-F(x1)
  else
    s2=lam4
 end
 s=[s1,s2]
endfunction

function [lam2,lam4,x0,x1,mu0,mu1,w0,w1]=fun(X)
 x=X(1);xd=X(2);y=X(3);yd=X(4);th=X(5);thd=X(6);xsi=X(7);eta=X(8);
 x0=x+xsi*sin(th)-l2*cos(th)
 x1=x+eta*sin(th)+l2*cos(th)
 mu0=Fx(x0)*(xsi*cos(th)+l2*sin(th))-xsi*sin(th)+l2*cos(th)
 mu1=Fx(x1)*(eta*cos(th)-l2*sin(th))-eta*sin(th)-l2*cos(th)
 w0=Fx(x0)*sin(th)+cos(th)
 w1=Fx(x1)*sin(th)+cos(th)
 lam2=(Fx(x0)*xd-w0*k*(xsi-d)/gam+mu0*thd-yd)*gam/w0^2
 lam4=(Fx(x1)*xd-w1*k*(eta-d)/gam+mu1*thd-yd)*gam/w1^2
endfunction
```

The road profile (the function $F(x)$ and its derivative) are defined below. The parameters a, b, f are available to vary the road surface.

```
function y=F(x)
  if x<0  then
   y=-a*cos(f*x)
  else
   y=a*(-2+cos(b*f*x))
```

```
  end
endfunction

function y=Fx(x)
  if x<0  then
    y=a*f*sin(f*x)
  else
    y=-a*f*b*sin(b*f*x)
  end
endfunction
```

We can avoid having to define $F_x(x)$ by using automatic differentiation, which is available as a plug-in toolbox for Scilab.

5.2.5 Simulation Result

The simulation is run with the following Scilab script:

```
A=30;a=5.2;f=%pi/A;B=3;b=1;                          ← curve parameters
XI=A;                                  ← used to define boundaries of display
del=.02;TT=[del:del:15];                      ← instances for animation
// model parameters
l=6;l2=l/2;m=1;d=4;g=10;gam=.5;
J=m*l2*l2/4;k=12;r2=d/4;
// compute reasonable initial state
x=-XI+.7;
y=d-m*g/2/k+F(x);xsi=y-F(x-l2)-0.3;eta=y-F(x+l2)-0.4;
X=[x;0;y;0;0;0;xsi;eta];
ST1=1;ST0=1;
[XX,T]=simul_car(TT,X);                              ← simulation
play(T,X,XI)                                         ← animation
```

The initial condition is selected so that the car starts off with zero speed close to the top of a hill. To avoid having to compute an initial condition in which both wheels are on the ground, we start off with both wheels slightly above the ground. The car lands and start rolling immediately.

The animation is performed by the function **play**, which can be found in Appendix B.2. During this animation, the system changes state a number of times. Some of the changes can be seen in Figure 5.4, which shows the position of the car at certain points during the simulation.

To make the animation more realistic, wheels are drawn even though the model assumed zero-sized wheels. It turns out that the simulation is valid (assuming of course that wheels have zero mass) if the function $F(x)$ describes not the profile of the ground but a function drawn by the center of a wheel if the wheel would roll on the ground. This is true provided the radius of curvature of the curve $F(x)$ is always larger than the radius of the wheel.

In the function **play**, the profile of the ground drawn (see Figure 5.4) is not $F(x)$, but it is obtained from $F(x)$ and $F_x(x)$, and, of course, the radius of the wheel. This is done for a set of x's to allow plotting the curve with enough detail. The mathematical expression of this curve is in general difficult to obtain.

Figure 5.4. Snapshots of the car during animation.

Note also that in `play` we use the `xor` mode to erase previous positions of the car. In the `xor` mode, drawing an object for a second time erases it. This provides a very efficient way of moving objects, in particular when some objects (for example the road in this case) do not move because these latter objects need not be redrawn.

5.3 Open-Loop Control to Swing Up a Pendulum

We consider here the classical control problem of erecting a free swinging pendulum. The set up consists of a pendulum mounted on a motor-driven cart traveling horizontally.

5.3.1 Model

The model is a special case of the model considered later in the part dedicated to Scicos; see Figure 11.13 and equations (11.3)–(11.4). Here we consider the case that the angle ϕ is zero. The model is then

$$(M + m)\ddot{z} + ml\ddot{\theta}\cos(\theta) - ml\dot{\theta}^2\sin(\theta) = u(t), \tag{5.36}$$

$$l\ddot{\theta} + \ddot{z}\cos(\theta) - g\sin(\theta) = 0. \tag{5.37}$$

The state of this system can be defined as follows:

$$x = \begin{pmatrix} z \\ \theta \\ \dot{z} \\ \dot{\theta} \end{pmatrix}. \tag{5.38}$$

The system dynamics can then be expressed as follows:

$$\dot{x} = f(x, u), \tag{5.39}$$

where f is easily obtained from (5.36)–(5.37).

5.3.2 Control Problem Formulation

The control problem considered is that of finding an input $u(t)$ to move the pendulum from an initial position, hanging at rest with the cart at the origin, to a final position with the pendulum upright with the cart restored to its original position. We look only for an open-loop control; in a real application such a control must be supplemented with a feedback control to stabilize the system around the open-loop trajectory.

The solution to this open-loop control problem is not unique: there are many control functions $u(t)$ over any interval $[0, T]$ that would move the state $x(0) = [0, -\pi, 0, 0]$ to $x(T) = [0, 0, 0, 0]$. Clearly there is a compromise between the interval length T and the size of u: the smaller T is, the bigger u is going to be.

A common formulation is to consider u to be bounded:

$$|u(t)| \leq U_{\max}, \ 0 \leq t \leq T,$$

and look for the shortest control interval (smallest T). In this formulation, since u enters the system equations linearly and there is no cost associated with it, from the optimality principle, it can be shown that the optimal open-loop control is bang-bang. This means that the optimal open-loop trajectory consists of a piecewise constant function taking values $+U_{\max}$ and $-U_{\max}$. So the control starts off with $u(t)$ equal to $+U_{\max}$ for a period of τ_1 (which could be zero). This pushes the state to x_1. Then the control switches to $-U_{\max}$ and remains there over a period τ_2; at the end of this period the state is at x_2, and so on until the end of the interval T is reached. See Figure 5.5.

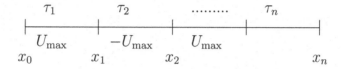

Figure 5.5. Controls and states over the interval $[0, T]$.

So the problem reduces to finding the number of "switchings" and their "times." To do so, we fix the number of switchings and optimize over the switching times. We can run the optimization problem for different numbers of switchings. We shall see the consequence of overestimating this number in an example.

5.3.3 Optimization Problem

The control problem is formulated as an optimization problem by defining the following cost function:

$$J(\tau) = h(\tau) + x(h(\tau))^T W x(h(\tau)), \tag{5.40}$$

where

$$\tau = (\tau_1, \ldots, \tau_n) \tag{5.41}$$

and

$$h(\tau) = \sum_{i=1}^{n} \tau_i. \tag{5.42}$$

The problem is then to minimize J over the τ's subject to the dynamics of the system. We shall use `optim` to solve the problem.

The computation of J requires the computation of the final state x_n given the control. This can be done with the Scilab function `ode`. But `optim` also requires the gradient of J with respect to τ $(= (\tau_1, \tau_2, \ldots, \tau_n))$. The gradient can also be obtained using `ode` and the linearized model of the system.

The linearized model can readily be obtained from (5.36)–(5.37):

$$(M + m)\delta\ddot{z} - ml\ddot{\theta}\sin(\theta)\delta\theta + ml\cos(\theta)\delta\ddot{\theta} - ml\dot{\theta}^2\cos(\theta)\delta\theta - 2ml\dot{\theta}\sin(\theta)\delta\dot{\theta} = \delta u,$$
$$l\delta\ddot{\theta} + \cos(\theta)\delta\ddot{z} - \ddot{z}\sin(\theta)\delta\theta - g\cos(\theta)\delta\theta = 0.$$

By straightforward algebra, it can be shown that the linearized model can be expressed as follows:

$$\dot{\delta x} = A(\dot{x}, x, u)\delta x, \tag{5.43}$$

where

$$A(\dot{x}, x, u) = \begin{pmatrix} 0 & I \\ V^{-1}F & V^{-1}G \end{pmatrix} \tag{5.44}$$

where

$$F = \begin{pmatrix} 0 & ml\sin(\theta)\ddot{\theta} + ml\dot{\theta}^2\cos(\theta) \\ 0 & \ddot{z}\sin(\theta) + g\cos(\theta) \end{pmatrix}, \tag{5.45}$$

$$G = \begin{pmatrix} 0 & 2ml\dot{\theta}\sin(\theta) \\ 0 & 0 \end{pmatrix}, \tag{5.46}$$

$$V = \begin{pmatrix} M + m & ml\cos(\theta) \\ \cos(\theta) & l \end{pmatrix}. \tag{5.47}$$

Note that z, θ, and their derivatives are all functions of u.

Note also that

$$\frac{\partial J}{\partial \tau} = \frac{\partial h}{\partial \tau} + x(h)^T W \frac{\partial x(h)}{\partial \tau}. \tag{5.48}$$

$\frac{\partial h}{\partial \tau}$ is simply a vector of ones. $\frac{\partial x(h)}{\partial \tau}$ is the variation in the final state x_n due to a variation in τ. Thus we can compute it by computing δx_n due to variations $\delta\tau$.

Let $\Psi(t)$ be the fundamental solution of the linear system (5.43), i.e., the solution of the following equation:

$$\dot{\Psi} = A(\dot{x}, x, u)\Psi, \quad \Psi(0) = I. \tag{5.49}$$

It is then straightforward to see that

$$\delta x_1 = f(x_1, U_{max})\delta\tau_1, \tag{5.50}$$
$$\delta x_2 = \Psi(\tau_2)\delta x_1 + f(x_2, -U_{max})\delta\tau_2 = \Psi(\tau_2)f(x_1, U_{max})\delta\tau_1 + f(x_2, -U_{max})\delta\tau_2$$

$$= \left(\Psi(\tau_2)f(x_1, U_{max}) \; f(x_2, -U_{max})\right)\begin{pmatrix} \delta\tau_1 \\ \delta\tau_2 \end{pmatrix}. \tag{5.51}$$

By continuing the same way, we can show that for all $i > 0$,

$$\delta x_i = \mathcal{G}_i \begin{pmatrix} \delta\tau_1 \\ \vdots \\ \delta\tau_i \end{pmatrix} \tag{5.52}$$

where $\mathcal{G}$ is obtained from the following recursion:

$$\mathcal{G}_{i+1} = \left(\Psi(\tau_i)\mathcal{G}_i \; f(x_i, (-1)^{i+1}U_{max})\right), \quad \mathcal{G}_0 = [\,]. \tag{5.53}$$

So at the end we have

$$\delta x_n = \mathcal{G}_n\delta\tau, \tag{5.54}$$

which allows us to evaluate $\frac{\partial J}{\partial \tau}$ thanks to (5.48):

$$\frac{\partial J}{\partial \tau} = \mathbf{1} + \mathbf{x_n^T W}\mathcal{G_n}, \tag{5.55}$$

where $\mathbf{1}$ denotes a row vector of ones of size n.

5.3.4 Implementation in Scilab

As we have seen in the previous section, the evaluation of the cost and the gradient of
the cost require the solution of the system and the fundamental solution of the linearized
system. These two systems cannot be solved separately because the linear system depends
on the trajectory of the original system. The two systems are thus solved simultaneously
by considering as state

$$X = \begin{pmatrix} x & \Psi \end{pmatrix}. \tag{5.56}$$

This matrix differential equation can be solved directly by `ode`. The associated Scilab
functions are defined in the following file:

```
function Xd=simul(t,X)
 x=X(:,1);dx=X(:,2:$)
 [xd,dxd]=full_model(x,dx)
 Xd=[xd,dxd];
endfunction

function [xd,dxd]=full_model(x,dx)
 [lhs,rhs]=argn(0);
 z=x(1);th=x(2);zd=x(3);thd=x(4);
 Mati=inv([M+m,m*l*cos(th);cos(th),l]);
 ydd=Mati*[m*l*sin(th)*thd^2+U;g*sin(th)];
 zdd=ydd(1);thdd=ydd(2);
 xd=[zd;thd;zdd;thdd];
 F=[0,m*l*thdd*sin(th)+m*l*thd^2*cos(th);0,zdd*sin(th)+g*cos(th)];
 G=[0,2*m*l*thd*sin(th);0,0];
 if rhs == 2  then dxd=[zeros(2,2),eye(2,2);Mati*F,Mati*G]*dx; end
endfunction
```

The cost and its gradient, to be used by `optim`, are then computed by the following
function

```
function [f,gg,ind]=cost(Ti,ind)
 xc=x0;
 dxdt=[];
 N=size(Ti,2)
 tc=0;
 for i=1:N
  U=(-1)^(i-1)*Umax;
  if Ti(i)>1d-9 then
    XX=ode([xc,eye(4,4)],tc,tc+Ti(i),1d-11,1d-12,simul)
  else
    XX=[xc,eye(4,4)]
  end
  xc=XX(:,1);tc=Ti(i)+Ti(i)
  Psi=XX(:,2:$);
  dxdt=[Psi*dxdt,full_model(xc)];
 end
 f=sum(Ti)+xc'*W*xc/2
 gg=ones(Ti)+xc'*W*dxdt
endfunction
```

The main Scilab script which uses optim is the following

```
⊢→ W=100*eye(4,4); // weighing matrix
⊢→ M=1;m=.2;l=.6;g=10; // model parameters
⊢→ N=5; // number of tau's
⊢→ delt=1.7;T0=ones(1,N)*delt/N; // initial guess
⊢→ x0=[0;-%pi;0;0]; // intial state
⊢→ Umax=9.7;
⊢→ [J,Topt,gr]=optim(cost,'b',ones(T0)*delt*0,delt*ones(T0),T
       0,'ar',300);

⊢→ Topt
 Topt  =

 !   0.    0.2817765    0.6133068    0.5066828    0.1752807  !
```

Note that we have let n be 5 (maximum of four switchings) but the optimal solution obtained by optim sets τ_1 to zero (Scilab variable Topt contains the vector τ). So this solution starts off with $u = U_{max}$ and switches three times over the interval. Both the simulation result and the animation show that the solution is correct; see Figures 5.6 and 5.7.

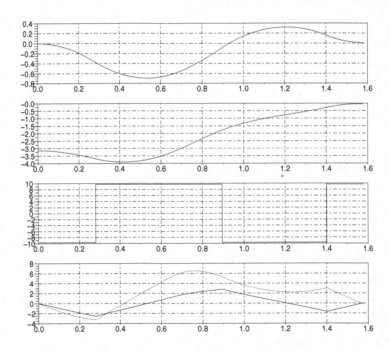

Figure 5.6. The first two figures illustrate the evolution of z and θ. The third figure is the control. The last figure contains the derivatives of z and θ showing that the pendulum comes to rest at the end of the interval as required.

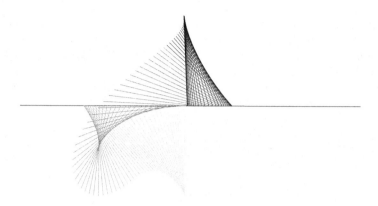

Figure 5.7. Snapshots of the pendulum during animation.

5.4 Parameter Fitting and Implicit Models

In many applications the scientist knows the general form of the equations that model the process, but specific parameter values are not known. Also, in many application areas the models are often differential algebraic equations (DAE). In this section we shall give an illustration of parameter fitting with a DAE model.

The general procedure will be the following. Suppose that the model that is to be found is a DAE

$$F(\dot{y}, y, t, p) = 0, \tag{5.57}$$

where p is a vector of unknown parameters. We suppose that the model is to be defined over the time interval $[0, T]$ and suppose that measurements are taken of the actual physical process for some function of y at times t_i for $i = 0, \ldots, N$. That is, we have the measured quantities

$$z_i = h(y(t_i), t_i), \quad i = 0, \ldots, N. \tag{5.58}$$

We suppose that J of these trials are done and let $z_{i,j}$ be the result of the jth trial at time t_i.

The initial condition for (5.57) could be fixed or it could be viewed as more parameters to be determined. We shall take the initial condition fixed, but that does not change the procedure. We wish to determine p such that the solution of (5.57) fits the observation data as closely as possible. There are a couple of ways to do this in Scilab. One way would be to use `optim`. Another is to use the function `datafit`. The function `datafit` will fit a function $e = G(p, z)$ to data. We found that `datafit` could solve the problem, but convergence could be slow and the accuracy was somewhat lower. Here we shall use directly `optim` with a `cost` function of the same form that `datafit` uses except that by using `optim` we can control the step used for the finite difference computation of the gradients.

We do not want to call for the integration of (5.57) for every z_{ij}. It is more efficient to do just one integration for each value of j. Thus we will take the z in $G(p, z)$ to be the vector of all $z_{i,j}$ for a fixed j.

5.4.1 Mathematical Model

Many applications use polynomial models. In some cases these are argued for on first principles. In other applications they occur because if linear equations do not suffice, then the next natural thing to try is a polynomial equation. If a linear approximation is not adequate, then the next level of Taylor approximation uses quadratic functions. The DAE we consider arises as the model of a batch reactor and is sometimes used as a test problem in the chemical engineering literature [19, 37]. The equations are

$$\dot{y}_1 = -p_3 y_2 y_8, \tag{5.59a}$$

$$\dot{y}_2 = -p_1 y_2 y_6 + p_2 y_{10} - p_3 y_2 y_8, \tag{5.59b}$$

$$\dot{y}_3 = p_3 y_2 y_8 + p_4 y_4 y_6 - p_5 y_9, \tag{5.59c}$$

$$\dot{y}_4 = -p_4 y_4 y_6 + p_5 y_9, \tag{5.59d}$$

$$\dot{y}_5 = p_1 y_2 y_6 - p_2 y_{10}, \tag{5.59e}$$

$$\dot{y}_6 = -p_1 y_2 y_6 - p_4 y_4 y_6 + p_2 y_{10} + p_5 y_9 \tag{5.59f}$$

$$0 = -0.0131 + y_6 + y_8 + y_9 + y_{10} - y_7, \tag{5.59g}$$

$$0 = p_7 y_1 - y_8(p_7 + y_7), \tag{5.59h}$$

$$0 = p_8 y_3 - y_9(p_8 + y_7), \tag{5.59i}$$

$$0 = p_6 y_5 - y_{10}(p_6 + y_7). \tag{5.59j}$$

In its original form this problem is known to be very stiff and is used for a test problem for parameter sensitivity software. We shall take a slightly modified set of initial conditions.

5.4.2 Scilab Implementation

To illustrate the procedure we shall proceed as follows. Our observational data will be given by a simulation followed by the addition of random perturbations. We begin with a fixed value of p, which we denote by `ptrue`. Assuming that `p` is given in the current Scilab environment, the system is coded as follows:

```
function [r,ires]=reactor(t,y,ydot)
  ires=0;
  r(1)=ydot(1)+p(3)*y(2)*y(8);
  r(2)=ydot(2)+p(1)*y(2)*y(6)-p(2)*y(10)+p(3)*y(2)*y(8);
  r(3)=ydot(3)-p(3)*y(2)*y(8)-p(4)*y(4)*y(6)+p(5)*y(9);
  r(4)=ydot(4)+p(4)*y(4)*y(6)-p(5)*y(9);
  r(5)=ydot(5)-p(1)*y(2)*y(6)+p(2)*y(10);
  r(6)=ydot(6)+p(1)*y(2)*y(6)+p(4)*y(4)*y(6)-p(2)*y(10)-p(5)*y(9);
  r(7)=-0.0131+y(6)+y(8)+y(9)+y(10)-y(7);
  r(8)=p(7)*y(1)-y(8)*(p(7)+y(7));
  r(9)=p(8)*y(3)-y(9)*(p(8)+y(7));
  r(10)=p(6)*y(5)-y(10)*(p(6)+y(7));
endfunction
```

Since the model (5.59) is a DAE, only some initial conditions will be consistent. We shall assume that $\{p_6, p_7, p_8\}$ are known and fixed at the values `ptrue(6:8)`. In this case the space of consistent initial conditions is independent of the remaining values of p_i and we do not have to worry about finding new consistent initial conditions as the parameters are varied.

We take fixed initial values of the first 6 components of the initial condition $y(0)$. The remaining values of $y(0)$ are found by solving the last four equations in (5.59) using `fsolve`. This ensures that our $y(0)$ is consistent to high order.

```
⟼ ptrue=[21.893;2.14e3;32.318;21.893;1.07e3;6.65e-3;4.03e-4;5.e-5/32];
```

```
⟼ function z=constraints(y2)          ← compute constraints for y=[y1,y2]
⟼    r = reactor(0,[y1;y2],0*ones(10,1));
⟼    z = r(7:10);
⟼ endfunction
```

```
⟼ function [y,v]=initial_condition(y1)    ← compatible initial value given y1
⟼    [y2,v]=fsolve(zeros(4,1),constraints);
⟼    y =[y1;y2];
⟼ endfunction
```

```
⟼ y01=[1.5776;8.32;0;0;0;0];          ← initial value of first 6 components
⟼ p=ptrue;
⟼ [y0,v]=initial_condition(y01);        ← a consistent initial condition
⟼ norm(v)                                ← test consistency
  ans  =
```

```
    2.262D-17
```

Simulations with `dassl` using $y(0)$ and `ptrue` show that the solutions are close to steady state by around $t = 0.5$. Accordingly we will take our observations on the interval $[0, 0.5]$. These simulations, which are indicated by the solid unmarked lines in Figs. 5.8–5.13, also show that the components of y_i vary greatly in size. A given experiment will consist of observing the value of the first six components of y every 0.05 seconds for a total of 10 time points. Hence the observation times are `tts=0.05:0.05:0.5`. We generate the experimental outcomes by taking a simulation using `ptrue` and then perturbing the entries by a uniformly generated random vector. The perturbations are scaled so that they are of the same relative size for each entry and each observation time. We take 9 such observations so that $J = 9$.

```
⟼ t0=0; tt=0:0.05:0.5;
⟼ y=dassl(y0,t0,tt,reactor);            ← integration with p=ptrue
⟼ y=y(2:7,2:11);y(4:6,:)=100*y(4:6,:);    ← rescale y(4:6,:)
⟼ rtrue=y(:);
```

```
⟼ rand('seed',20);
⟼ N=9;                              ← generate 9 observations perturbed with noise
⟼ obs= rtrue*ones(1,N);
⟼ obs= obs.*(1+ 0.1*(rand(60,N,'u')-0.5));    ← perturbations are to be rela-
                                           ← tive to the size of what is being perturbed
```

We then make an intial guess `p0` for the first 5 parameters and use `optim` to find the values of $\{p_1, \ldots, p_5\}$ to best fit the data. The final estimate is denoted by `pest`. This last step is carried out now by

```
⟼ function e=G(p1)                  ← error function to be minimized
```

```
⟼    p=[p1;ptrue(6:$)]                                    ← p(6:8) is fixed and fit using p(1:5)
⟼    rr=dassl(y0,t0,tt,1.e-10,1.e-12,reactor);
⟼    rr=rr(2:7,2:11);rr(4:6,:)=100*rr(4:6,:);
⟼    rr=rr(:);
⟼    e=0;
⟼     for i=1:size(obs,'c')
⟼       gg = (rr-obs(:,i));
⟼       e = e + gg'*gg;
⟼     end
⟼   endfunction


⟼   function [f,g,ind]=costf(p,ind)                       ← cost function for optim
⟼     if ind==2|ind==4  then
⟼        f=G(p);
⟼     else
⟼        f=0;
⟼     end;
⟼     if ind==3|ind==4  then
⟼        g=0*p;
⟼        pa=sqrt(%eps)*10*abs(p);
⟼        f=G(p);
⟼         for j=1:size(p,'*') do
⟼          v=0*p;v(j)=pa(j),
⟼            g(j)=(G(p+v)-f)/v(j);
⟼         end;
⟼     else
⟼        g=0*p;
⟼     end
⟼   endfunction


⟼ p0=ptrue(1:5)*0.5;                                      ← initial guess for p
⟼ [err,pest]=optim(costf,p0,'ar',100,100,1.e-9,1.e-9,imp=2);


⟼ [p0,pest,ptrue(1:5)]                                    ← fitted parameters
  ans  =

!   10.9465     10.614308    21.893 !
!   1070.       1105.6571    2140.  !
!   16.159      33.346866    32.318 !
!   10.9465     5.8412889    21.893 !
!   535.        946.85234    1070.  !


⟼ [G(p0),G(pest),G(ptrue(1:5))]                           ← cost minimized by datafit
  ans  =

!   453.75872    6.1442786    5.2443056 !
```

The first thing one notices is that `pest` is not the same as `ptrue(1:5)`, nor is it even very close. However, if we examine the answer we see that we have in fact found a good approximation. Running the following script will graph the different y_i for `p0`, `ptrue`, and `pest`. What we see is that the new value `pest` is providing a good fit to the data. This example shows the importance of using parameter bounds whenever they are available especially in nonlinear problems. Observe how the fifth component of `pest` has moved much farther from the fifth component of `ptrue` than the fifth component of `p0` was.

```
p=[ptrue]; krt=dassl(y0,t0,tt,reactor);
p=[p0;ptrue(6:$)];kr0=dassl(y0,t0,tt,reactor);
ttt=[0.01:0.02:0.5];
p=[pest;ptrue(6:$)];krf=dassl(y0,t0,ttt,reactor);
scale_r=[ones(1,3),100*ones(1,3)];
 for ii=2:11
  f1=scf(ii-1);
  plot2d(ttt,krf(ii,:));
  plot2d(ttt,krf(ii,:),-1);
  plot2d(tt,krt(ii,:));
  plot2d(tt,kr0(ii,:));
  plot2d(tt,kr0(ii,:),-2);
  xtitle('y('+string(ii-1)+')');
  nz=size(obs,'c');
   if nz <>1 & ii<=7     then
     for k=1:N
       obsk=obs(:,k);obsk=matrix(obsk,6,10);
       plot2d(tt(2:$)',obsk(ii-1,:)./scale_r(ii-1),-3);
     end
   end
 end
```

Figures 5.8–5.13 illustrate what happens for several components for y. In all cases the solid unmarked line gives the graph of y_i using `ptrue`. The graphs using `pest` and `p0` are indicated by $+$ and $\times$'s respectively. In the cases where there are observations, those are indicated by circles.

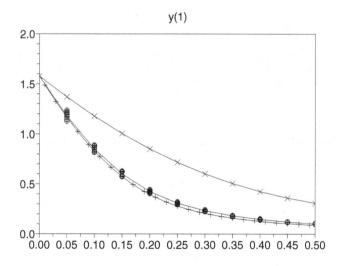

Figure 5.8. Graph of observational data and y_1 for `ptrue` (solid), `pest` ($+++$), and `p0` (xxx).

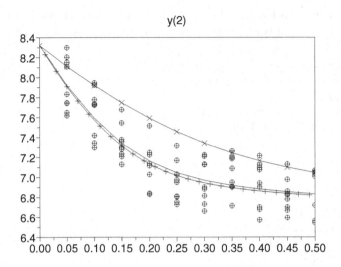

Figure 5.9. Graph of observational data and y_2 for **ptrue** (solid), **pest** $(+++)$, and p0 (xxx).

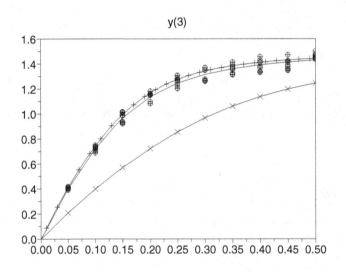

Figure 5.10. Graph of observational data y_3 for **ptrue** (solid), **pest** $(+++)$, and p0 (xxx).

Often there are bounds on reasonable parameter values which come either from physical arguments or practical experience. In this case the bounds can be incorporated into the call for **datafit** or **optim**. Suppose, for example, that we knew that the value of p_i was in the interval $[0.8 * \text{ptrue}(i), 1.2 * \text{ptrue}(i)]$. If we replace the last line of the first script with

```
ps=ptrue(1:5);pinf=ps;psup=1.2*ps;p0=1.1*ps;
[err,pest]=optim(costf,'b',pinf,psup,p0,'ar',100,100,1.e-9,1.e-9,imp=2);
```

we get the new parameter estimate of

```
[p0,pest,ps]
```

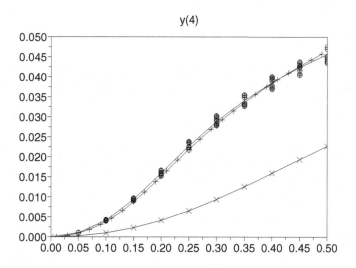

Figure 5.11. Graph of observational and y_4 for **ptrue** (solid), **pest** (+++), and p0 (xxx).

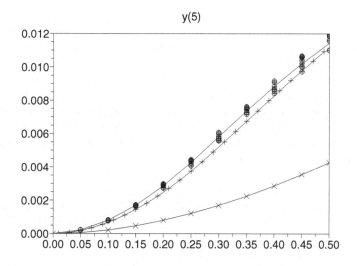

Figure 5.12. Graph of observational data and y_5 for **ptrue** (solid), **pest** (+++), and p0 (xxx).

```
ans   =

!    10.9465      24.172324     21.893 !
!    1070.        2349.0207     2140.  !
!    16.159       32.318        32.318 !
!    10.9465      24.364568     21.893 !
!    535.         1083.1877     1070.  !
```

y(6)

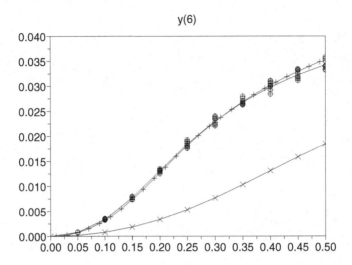

Figure 5.13. Graph of observational data and y_6 for `ptrue` (solid), `pest` (+++), and p0 (xxx).

Comments

`optim` calls an iterative solver. Each time G is evaluated there is a call to `dassl`. For more complicated problems with more parameters or longer time intervals these calls can be computationally expensive. As implemented in the previous illustration there would be N calls for each value of p. If these calls are taking too much time, one can directly formulate the cost as $\sum_i e_i^T e_i$, use `optim`, and have to do only one call to `dassl` for each value of p.

Parameter estimation is an important part of many engineering and scientific procedures. The example given illustrates two important facts. One is that since one is using the data to construct a model that predicts the data, it is important to have enough data to cover the region in which one wishes to use the model. Secondly, the criterion is the ability to predict the observations. If the parameter estimates are of interest themselves, then considerable more analysis and care may be needed to determine how good the estimate is.

Notice in this example that the graphs using the estimated parameters appear to be close to the graphs for the true parameter values. So the numerical algorithm is doing a good job of giving us parameter values that give the desired behavior. However, the estimated parameter values are not always close to the true parameter values. A similar behavior was observed in Section 3.4 in the discussion of `lsqrsolve`.

In the nonlinear case there are the added problems of local minima when one is solving for parameter estimates. But even for the linear case $y = Ax$, it is important to keep in mind that the following three problems are distinct and can produce different estimates $\hat{A}$ for A:

1. Given data $\{x_i, y_i\}$ find the value $\hat{A}$ that gives the best prediction law for y given an additional value of x.
2. Given data $\{x_i, y_i\}$ find the value $\hat{A}$ that gives the best prediction law for x given an additional value of y.
3. Given data $\{x_i, y_i\}$, which we assume come from a model $y = Ax$, find the value of $\hat{A}$ that gives the best estimate of A.

The functions `optim`, `datafit`, and `lsqrsolve` as described here are being used for the first type of problem since the error in the value of G is being minimized. The function G needs to be formulated differently if either problem 2 or 3 is to be solved.

Finally, this example illustrates an important aspect of solving optimization problems. Finding the optimal solution even when we know it exists can be difficult. When we initially solved this problem we used more of the default settings of `optim` and `dassl`. The result was not nearly as good as reported above. After some experimentation, it was determined that one factor was that the gradient approximation was not accurate enough. Since the function being differenced was the result of an integration by `dassl`, we both increased the lower bound on the difference in `optim` and tightened the error tolerances in `dassl`. Another problem was that the large difference in magnitude of the components was leading to inaccuracy. So we put a factor of 100 in the function evaluation, which helps prevent stagnation and increased accuracy of gradients. In particular, we rescaled `y(4:6)` to give these variables a range similar to those in `y(1:3)`. This was necessary to increase the part dedicated to `y(1:3)` in the global cost.

We also gave `optim` some guidance. In

```
[err,pest]=optim(costf,p0,'ar',100,100,1.e-9,1.e-9,imp=2)
```

the `1.e-9` sets the thresholds on the gradient norm and the decrease in `f`. Here we are forcing the change in `f` and the gradient to be rather small at termination since they need to be below these thresholds. The `imp=2` gives a verbose reporting on the iterations, which was used in examining the behavior of the iteration while deciding where to set the thresholds. In trying to solve a parameter problem in practice when the solution is not known it is often helpful to try first what we have done here. That is, generate observations by simulation with known parameters and then try first to find the known parameters. This helps in checking that the problem is properly coded and gives insights into the problem's behavior. Thus the problem solved in this section is not as academic as it appears and is representative of part of the solution of real problems.

## Part II

Scicos

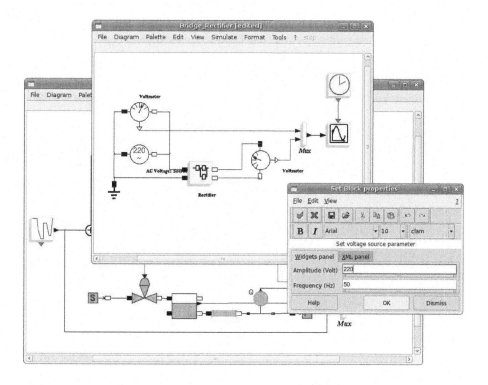

6

Introduction

Scicos is a ScicosLab toolbox for modeling and simulation of dynamical systems. Scicos is particularly useful for modeling systems where continuous-time and discrete-time components are interconnected. Such models can be programmed directly in Scilab using the `ode` and `dassl` functions, studied previously, but that requires programming the discrete-time dynamics in the Scilab language. These programs are often complex and difficult to debug. Moreover, there is no simple systematic way of programming them in a modular fashion.

Scicos provides a modular way to construct complex dynamical systems using a block diagram editor. Scicos diagrams are compiled and simulated efficiently by a single click. Scicos handles, in particular, the interaction between continuous-time dynamics and system events including events associated with the timing of a discrete-time clock. Such events affect the way the numerical solver, which integrates the continuous-time dynamics, should be called. Handling efficiently such matters by hand for complex dynamical systems can be extremely difficult.

Using Scicos, the user can construct a library of reusable modules (blocks) that can be used in different models in different projects. This is particularly useful when a large model is composed of modules designed by different development teams. The Scicos formalism, which must be respected in designing blocks and submodels, guarantees that modules constructed separately and interconnected can work harmoniously together.

A large number of blocks are already available in Scicos palettes. These blocks provide elementary operations needed to construct models of many dynamical systems. Users seldom need to construct a new block from scratch.

Scicos is more than a modeling and simulation program. It contains many functionalities to help the designer optimize model parameters, validate models, generate C code, etc. In this part of the book, we present Scicos and some of these useful functionalities. We shall give the reader a general picture of the Scicos environment and the type of applications that can be treated in it. Numerous examples are used to illustrate the subjects covered. Enough information is provided to allow the user to understand the examples. The examples are also included in the book's web site.

We start by providing a tutorial in Chapter 7 on how to use the editor to construct models and how to simulate these models. We cover important editor functionalities and illustrate them using simple examples.

Chapter 8 is devoted to a description of the formalism behind Scicos. This chapter provides detailed information on how blocks are activated by activation signals, how data is exchanged between blocks, and what synchronism means in this environment and how it can be used for conditional subsampling.

S.L. Campbell et al., *Modeling and Simulation in Scilab/Scicos with ScicosLab 4.4*,
DOI 10.1007/978-1-4419-5527-2_6, © Springer Science+Business Media, LLC 2010

Even if in many cases Scicos diagrams can be constructed using existing blocks, sometimes it is more convenient, or even becomes necessary, to develop new blocks. This is the subject of Chapter 9. Sometimes a new block is not absolutely necessary but can simplify the diagram and/or its use can make the simulation of the diagram more efficient. Various ways of constructing new blocks are discussed in this chapter. The content of this chapter, however, is fairly technical, and most users may skip it on a first reading.

In Chapter 10, through a number of examples, we show what types of systems can be modeled in Scicos and propose solutions for modeling certain behaviors that are particularly difficult to realize outside of Scicos.

Often the objective for modeling and simulation is parameter optimization. The parameter can be a system parameter to be tuned in an identification process, or a controller or filter parameter to be adjusted to a given system. The procedure for parameter optimization consists in associating a cost (obtained by simulation) to a set of parameters. This cost is then optimized by standard optimization techniques. This procedure, which usually requires many simulation runs, can easily be implemented in the Scilab-Scicos environment by batch processing. In Chapter 11, we present Scilab functions used for batch simulation of Scicos diagrams in Scilab and present a number of examples illustrating typical applications.

Code generation is an important functionality in Scicos. A typical application is the situation in which the diagram models a controlled system. The control part of the diagram, after validation of the control law through simulation, is to be implemented in an embedded system. For that, C code generation can be performed for a part of the diagram designated by the user. Code generation is the subject of Chapter 12. Examples are provided and some pitfalls are discussed.

A number of debugging functionalities are available in Scicos. They are explained in Chapter 13. Debugging is particularly useful when diagrams become complex and when user-defined blocks are used. A few examples are provided to show how debugging can be done efficiently.

An important feature of Scicos is its capacity to use "implicit" blocks in a diagram. This allows the construction of systems using blocks modeling physical components in a natural way. These blocks have no explicit inputs and outputs but rather have ports. Connecting these ports by links defines constraints. For example, in this environment we can define an electrical circuit with implicit blocks modeling resistors, capacitors, diodes, and other electrical components, and with links imposing Kirchoff's laws of current and equality of voltages at the two connected ports. The possibility to mix implicit and normal blocks in the same environment provides a powerful modeling environment that is presented in Chapter 14.

Appendix A contains an introduction on how Scicos is implemented and, in particular, what data structures it uses. Understanding implementation issues is very useful not only for advanced users who want to customize the tool, but also for anybody who encounters subtle modeling problems and wants to take full advantage of debugging facilities. This chapter also gives a lot of information on where important Scicos programs can be found. Most of these programs (except for the simulator part) are Scilab programs, so they are easy to read and customize.

In this book we present functionalities of Scicos distributed with ScicosLab version 4.4 or higher.

Note also that running some of the examples presented here requires a C compiler. This is usually no problem under Linux, MacOSX, and most Unix workstations. But under the Windows operating system, it requires Visual C++, i.e., the installation of Visual Studio or Visual Studio Express (freely available from Microsoft). Visual C++ is automatically instantiated and usable by ScicosLab.

7

Getting Started

7.1 Construction of a Simple Diagram

Scicos contains a graphical editor that can be used to construct block diagram models of dynamical systems. The blocks can come from various palettes provided in Scicos or can be user-defined. In this section, we describe how the editor can be used to construct simple models and how these models can be simulated.

7.1.1 Running Scicos

Scicos is a Scilab toolbox included in the ScicosLab package. The Scicos editor can be opened by the `scicos` command

$\longmapsto$ `scicos;`

This command opens up an empty Scicos diagram named by default `Untitled`. Called with an argument, it can open up an existing diagram:

$\longmapsto$ `scicos my_diagram.cos;`

The Scicos main window displaying an empty diagram is illustrated in Figure 7.1. The look of the main window may be slightly different under different window managers and operating systems. Editor functionalities are available through pull-down menus placed at the top of the editor window. The manual page of each functionality is displayed by selecting `Help` in the `Misc` menu and then the item of interest in the menu.

Some of the editor functionalities can also be accessed by clicking the right mouse button. Finally, keyboard shortcuts can be used for various operations. For example, typing an "r" activates the operation `Replot`, which centers and redraws the diagram. Keyboard shortcuts can be defined by the user.

Note that when Scicos is running, the Scilab window is not active. It can be activated with the `Activate Scilab Window` button in the Scicos `Tools` menu. This operation does not close Scicos, it simply deactivates it temporarily. To reactivate Scicos, do a simple mouse click on any Scicos diagram, or execute the `scicos` command in Scilab.

7.1.2 Editing a Model

To construct a model, we need to access Scicos blocks. Scicos provides many elementary blocks organized in different palettes that can be accessed using the operation `Palettes`

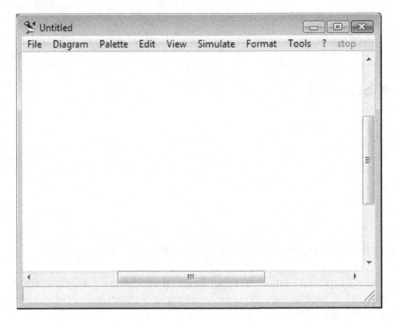

Figure 7.1. Scicos editor main window.

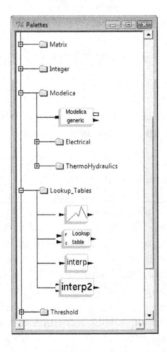

Figure 7.2. Palette browser.

in the **Edit** menu. This operation opens up a dialog box that includes the list of available palettes. By selecting a palette in the list, a new Scicos window opens up displaying the blocks available in this palette. A more direct and convenient way of accessing a block in a palette is via the **PalTree** operation in the **Edit** menu. This operation opens a Palette browser; see Figure 7.2. Blocks from palettes can be copied into the main Scicos window by dragging the desired block and dropping at the location where the block is to be copied in the Scicos window.

The **Sources** and **Sinks** palettes contain respectively blocks generating signals without any inputs and blocks without any outputs such as scopes and Write-to-file blocks. We start by copying from these palettes three blocks as illustrated in Figure 7.3. In this figure we have copied a sinusoid signal generator, a scope, and an event generator (**Event Clock**). The first block generates on its unique output port a sine function. We wish to display this signal using the scope. This can be done by connecting the output of the signal generator to the input of the scope by clicking first on the output port and then near the input port of the scope. The **Align** operation in the **Edit** menu can be used to align the two ports beforehand in order to make sure the link becomes horizontal.

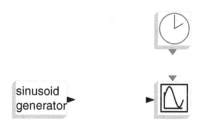

Figure 7.3. Scicos diagram in construction. Blocks are copied from the palettes.

The **Event Clock** is used to activate the scope block periodically with the desired frequency. Every time the scope is activated, it reads the value of the signal on its input port (which is nothing but the value of $\sin(t)$ generated by the signal generator). This value is then used to construct the curve that is displayed in the scope window. To specify that the scope block is activated by the event generator, the activation output of the event generator should be connected to the activation input of the scope. The result then looks like the diagram in Figure 7.4. This diagram is now complete.

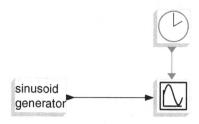

Figure 7.4. Completed Scicos diagram.

Note that Scicos diagrams contain two different types of links. The regular links transmit signals and the activation links transmit activation timing information. By default, the activation links are drawn in red and the regular links in black, but this can be changed by the user. Also note that regular ports are placed on the sides of the blocks whereas activation inputs and outputs are respectively on the top and at the bottom of the blocks. Most blocks follow this convention, but the user can define new blocks with ports arbitrarily placed.

The editor provides, through pull-down menus, many functionalities to change the look of the diagram and that of the blocks such as changing color, size, and font.

7.1.3 Diagram Simulation

To simulate a diagram, it suffices to select the Run operation from the Simulate menu. Simulation parameters can be set by the Setup operation in the same menu. There we can, for example, adjust the final simulation time.

Running the simulation for the system in Figure 7.4 leads to the opening of a graphics window and the display of a sinusoidal signal. This window is opened and updated by the scope block. The simulation result is given in Figure 7.5. In this case, the final simulation time is set to 30. The default value of the final simulation time is very large. A simulation can be stopped using the stop button on the main Scicos window, subsequent to which the user has the option of continuing the simulation, ending the simulation, or restarting it.

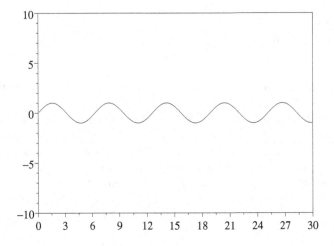

Figure 7.5. Scope window for the system in Figure 7.4.

A Scicos diagram can be modified and simulated again. Let us consider adding an integrator to the diagram in Figure 7.4 as illustrated in Figure 7.6. The integrator block gives as an output the integral of its input. This block comes from the Linear palette. We also replace the scope with a multi-input scope. Note that to create a split on a link, in particular here to create a link going to the integrator, off the link connecting the signal generator to the scope, the user must click first on the existing link at the position where the split is to be placed. Also note that to create a broken link, the user can click on intermediary points before clicking on the destination, which is necessarily an input port.

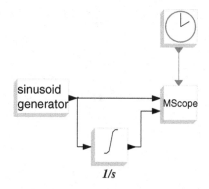

Figure 7.6. Modified Scicos diagram.

The simulation result for this new diagram is given in Figure 7.7.

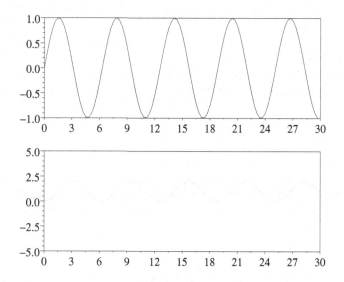

Figure 7.7. Scope window after simulation of the modified diagram.

7.1.4 Changing Block Parameters

The behavior of a Scicos block may depend on parameters that can be modified by the user. These parameters can be changed by clicking on the block. This action opens up a dialog box showing current values of the block parameters and allowing the user to change them. For example, the dialog box associated with the integrator block is illustrated in

Figure 7.8. The other blocks also have parameters. For example, in the sinusoid generator, we can set the frequency, the amplitude, and the phase. In the same way, certain properties of the scope window can be set by the parameters of the scope block.

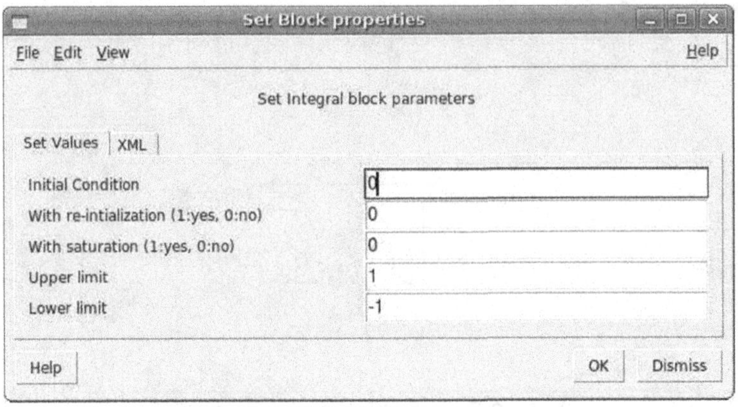

Figure 7.8. Dialog box of the integrator block.

The value of a parameter can be defined by any valid Scilab instruction. For example, the frequency of the sinusoid generator can be set to `2*%pi/10`. The Scilab instruction can also include Scilab variables, but these variables must have been previously defined in the "context" of the diagram. Such variables are called *symbolic parameters*. Symbolic parameters and the context will be discussed in the next section.

Let us now consider more complicated examples in which the systems we want to model are given by differential equations. An integrator block is used to define each state of the system.

SIR Model for Spread of Disease

We consider a simple model of how a disease can spread in a community [5]. Let $s(t)$ be the fraction of the population susceptible to getting infected as a function of time t. Also let $i(t)$ be the infected (and infectious) fraction of the population and $r(t)$ the recovered, and thus immune, fraction of the population. Then the SIR model can be expressed as follows:

$$\dot{s} = -as(t)i(t), \tag{7.1a}$$
$$\dot{i} = as(t)i(t) - bi(t), \tag{7.1b}$$
$$\dot{r} = bi(t), \tag{7.1c}$$

where a and b are positive parameters.

To model this system in Scicos, we set the state of one integrator to s, another one to i, and a third one to r. It is then straightforward to construct the inputs of the integrators from their outputs, as can be seen in Figure 7.9. Here we have set a to 1 and b to 0.3. For the reader less familiar with block diagrams, the $\times$ and Σ blocks multiply and add inputs respectively. Since we are calling the outputs of the integrator blocks s, i, and r, the Scicos implementation of the differential equation in (7.1a) can be thought of as the integral equation

$$s(t) = -\int_{t_0}^{t} s(\tau)i(\tau)d\tau + s(t_0),$$

where $s(t_0)$ is the initial condition of s at time t_0. Similarly (7.1a), (7.1c) are implemented as integral equations in the Scicos diagram. However, they are integrated as differential equations by Scicos using a numerical ODE solver.

The simulation result for this SIR model is given in Figure 7.10. The variable s is initialized to 0.999, and i to 0.001.

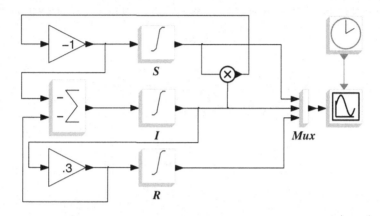

Figure 7.9. Scicos implementation of the SIR model.

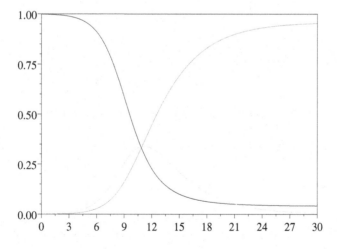

Figure 7.10. The simulation shows, as expected, that the percentage of the recovered population is increasing and that of the susceptible population is decreasing. The percentage of population infected reaches its maximum at a time called the peak of the epidemic.

Chaotic Dynamics of a Rössler Attractor

The Rössler system [54] given below has chaotic behavior for certain values of the parameters a, b and c:

$$\dot{x} = -(y + z),$$
$$\dot{y} = x + ay,$$
$$\dot{z} = b + z(x - c).$$

This system is modeled in Figure 7.11 with $a = b = 0.2$ and $c = 5.7$. The initial conditions are set to zero. The 2D scope is used to plot y against x. The result is given in Figure 7.12.

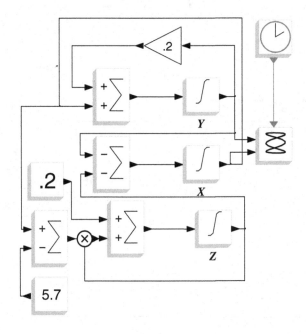

Figure 7.11. Scicos implementation of the Rössler attractor.

7.2 Symbolic Parameters and Context

Often it is useful to use symbolic parameters to define block parameters. This is particularly the case if the same parameter is used in more than one block or if the parameter is computed as a function of other parameters. Symbolic parameters are simply Scilab variables that must be defined in the context of the diagram before being used in the definition of block parameters.

Each Scicos diagram contains a context. The context is simply a Scilab script used to define Scilab variables, which can then be used to define block parameters. To access

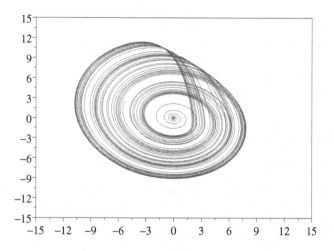

Figure 7.12. Output of the 2D scope for the Rössler model.

the context of the diagram, use the **Context** button in the **Edit** menu. This opens up an editor. The user can use this editor to write a Scilab script, that is, a set of Scilab commands that are executed by Scilab after a click on the **ok** button. For example, if the command **C=1** is placed in the context, then **C** can be used to define a block parameter. If the value of **C** is changed in the context, the parameter of the block is automatically updated accordingly.

Let us now consider an example in which the context is very useful for defining a generic diagram. We consider the construction of a sampled-data observer for a linear system. The system is considered to be modeled as a continuous-time state-space linear system:

$$\dot{x} = Ax + Bu, \tag{7.2}$$

$$y = Cx, \tag{7.3}$$

where A, B, C are constant matrices.

In developing controls for a system like this, it is often easier to develop them in terms of the state x. State feedback controls are one example. But the only thing that may be available in practice is the output y. A state observer is another dynamical system, which accepts as input u and y, and gives as an output $\hat{x}$, which has the property that $x - \hat{x}$ goes to zero independent of the initial conditions of either (7.2) or the observer. Observers have the nice property of being able to return to giving good estimates after disturbances. The speed with which the error goes to zero can be specified to give good estimation but avoid being overly sensitive to disturbances.

A continuous-time observer can be constructed as follows:

$$\dot{\hat{x}} = A\hat{x} + Bu + K(y - C\hat{x}). \tag{7.4}$$

The matrix K must be chosen so that the eigenvalues of $A - KC$ have negative real parts. This ensures that the estimation error $\tilde{x} = \hat{x} - x$ satisfying $\dot{\tilde{x}} = (A - KC)\tilde{x}$ will go to zero.

The discrete-time (sampled data) observer is obtained by first constructing the corresponding continuous-time observer using the method of pole placement, and then discretizing it. We begin by taking fixed dimensions and randomly generated matrices. We will then explain how to make this a generic diagram.

The Scilab script to perform this procedure with random matrices, placed in the context of the diagram, is the following:

```
n=5;m=2;dt=.2;
A=rand(n,n);A=A-max(real(spec(A)))*eye()
B=rand(n,1);C=rand(m,n);D=zeros(m,1);
x0=rand(n,1);
K=ppol(A',C',-ones(x0))';
Ctr=syslin('c',A-K*C,[B,K],eye(A),zeros([B,K]))
Ctrd=dscr(Ctr,dt)
[Ad,Bd,Cd,Dd]=abcd(Ctrd)
```

The original system is constructed with random matrices, and the A matrix is modified to make sure the system is not exponentially unstable. The Scilab pole placement (eigenvalue assignment) function ppol is used to obtain the observer gain matrix K, and dscr is used to discretize the observer. dt is the sampling period.

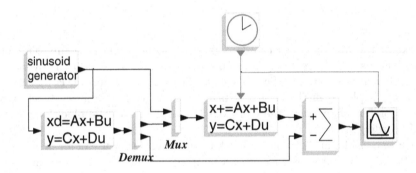

Figure 7.13. Scicos diagram including the system and the discrete-time observer.

The Scicos diagram in Figure 7.13 is used to simulate the observer performance. The input u is set to $\sin(t)$ and the estimation error is displayed using a scope. The continuous-time and discrete-time state-space linear system blocks are used to model the original system and the observer. The parameters of the first are set as illustrated in Figure 7.14 and the second as in Figure 7.15. The first block also outputs its internal state to be used for generating the estimation error displayed by the scope. dt is used to define the period of the Event Clock.

The use of symbolic parameters is particularly useful in this case because it allows us to construct a generic system observer diagram. To change the system matrices A, B, and C, the size of the system, the size of the outputs, or the discretization time, we only have to modify the definition of the m, n, and system matrices. No change is made to the diagram. In fact, by rewriting the context as follows we can make the diagram completely generic:

```
load('datafile')
K=ppol(A',C',-ones(x0))';
Ctr=syslin('c',A-K*C,[B,K],eye(A),zeros([B,K]))
Ctrd=dscr(Ctr,dt)
```

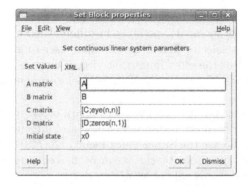

Figure 7.14. Dialog box of the block realizing the original system; the output of the block contains also its state.

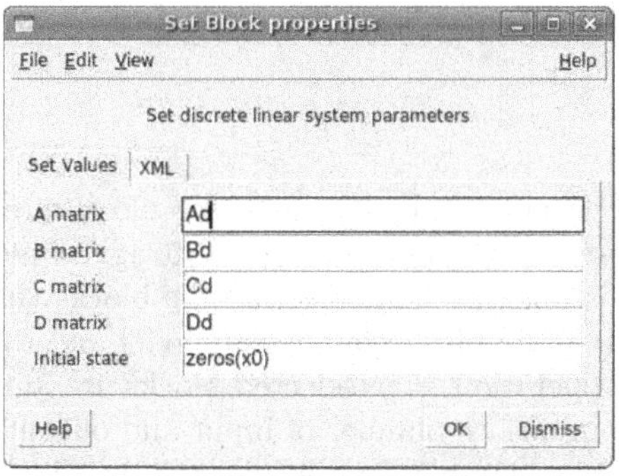

Figure 7.15. Dialog box of the discrete-time linear system realizing the observer.

```
[Ad,Bd,Cd,Dd]=abcd(Ctrd)
```

We just have to make sure that the data file `datafile` contains variables A, B, C, D, x0, and `dt` before launching Scicos. This file can be created in Scilab as follows:

$\longmapsto$ `save('datafile'',A,B,C,x0,dt)`

Remark

If the context contains many lines of Scilab code, it is convenient to place the code in a separate script file and execute it with a single **exec** command in the context. However, if the file being executed by the **exec** command is changed when the diagram is already open, and the user wants the modification to be taken into account, he or she should do an **Eval** because Scicos has no way of knowing whether the file has been changed.

It should also be noted that using a separate script file implies that the Scicos diagram is not self-contained and this script file must always accompany the diagram.

7.3 Virtual blocks and Hierarchy

Some blocks in Scicos palettes are not standard blocks realizing computations during the simulation. These virtual blocks facilitate the modeling task and are often used by the designer to improve the presentation of the Scicos diagram. For example the `Goto` and

`From` blocks can be used to create a link without having to draw it. This is useful when the link is connects two blocks far apart, possibly in different submodels. Similarly, the bus creator (`BUS`) and the bus selector `DEBUS` blocks allow user to group a set of links into a single "link" called bus. But the most important virtual block is a Super block, which will be presented in details here. Note that the virtual blocks are removed in the first phase of compilation.

It is not good modeling practice to place too many blocks in a diagram at the same level because the diagram becomes incomprehensible and difficult to read and debug. For large systems, it is useful to use the Super block facility to construct a hierarchical model. A Super block looks like any other block: it can be moved, copied, resized, etc., but its behavior is defined by the Scicos submodel within it.

7.3.1 Placing a Super Block in a Diagram

There are two ways to place a Super block in a Scicos diagram.

Region-to-Super-block

If the submodel the user wants to place in a Super block already exists in the model, then the `Region-to-Super-block` operation of the `Diagram` menu can be used to place it in a Super block. This is done by first selecting the blocks (usually done by specifying the region of the diagram) to be placed in the submodel and then the application of the `Region-to-Super-block` operation. The selected blocks are automatically replaced with a Super block with the appropriate number of input and output ports.

Let us consider a simple example. In the diagram of Figure 7.13, we used a linear system block to model the original system, but since we also needed its internal state, we had to concatenate it to the output, which we then had to split using the `Demux`[1] block in order to generate separately the state x and the output y. The diagram would have been much clearer if the linear system block had two output ports generating x and y. Unfortunately, that is not the way this block works.

But the combination of the linear system block and `Demux` does exactly what we want. So it is natural to consider constructing a Super block out of these two blocks. This can be done using the `Region-to-Super-block` functionality and selecting a region containing these two blocks. The result is illustrated in Figure 7.16.

Super Block in Palette

A Super block can also be placed in a diagram by copying the empty Super block provided in the `Others` palette. The desired submodel can then be constructed inside the Super block.

7.3.2 Editing a Super Block

To edit the content of a Super block, it suffices to click on it to "open" it. This opens up a new Scicos editor window resembling the main Scicos window displaying the content of the Super block. For example, by opening the Super block in Figure 7.16, the model illustrated in Figure 7.17 is displayed in the new editor.

---

[1] Demux is network terminology for demultiplex and it means to separate channels.

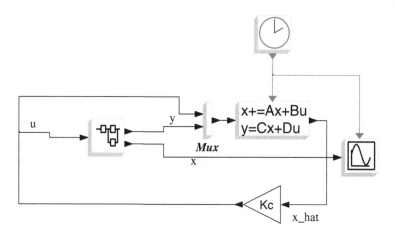

Figure 7.16. The linear system and `Demux` blocks are placed in a Super block.

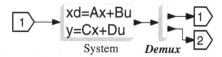

Figure 7.17. Inside the Super block.

Note that new blocks have been added to the submodel that were not present in the original model. These are the input, output port blocks. These blocks define the connection from the inside of a Super block to the outside world. There are exactly as many input (respectively output) port blocks inside the Super block as there are input (respectively output) ports on the Super block.

The Super block can be edited just like the main Scicos diagram. If the number of input or output port blocks is modified, the number of ports of the Super block in the main window is adjusted automatically accordingly. Note that the port blocks must be numbered consecutively and the corresponding ports on the Super block are also numbered counting consecutively from the top to the bottom.

Any number of Super blocks, at various levels of hierarchy can be open and edited in the Scicos editor at any time. Copy and paste operations work across the corresponding editor windows.

There is no need to close any diagram to compile and run simulations; simply use the `Simulation` menu on the top diagram while leaving all other windows open without any modification. There is also no need to close any windows to access the Scilab shell at prompt level. For that use the `Activate Scilab Window` button in the `Tools` menu. This is particularly useful for loading/defining new functions (for example new block interfacing functions), performing computations (post/pre data processing) using From and To Workspace blocks, etc. We can return to Scicos by simply clicking on any Scicos window.

The hierarchical structure of a Scicos diagram can be visulized in a browser using the `Browser` button in inside the `Tools` menu. This browser can also be used to open any diagram simply by double-clicking on its name. See Figure 7.18. The current diagram (the one that has focus) is highlited in the browser.

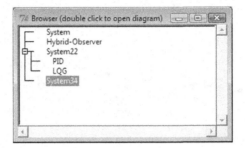

Figure 7.18. Scicos browser.

7.3.3 Scope of Variables in Super Block Contexts and Masking

As we have seen previously, block symbolic parameters are Scilab variables that are defined in the context of the diagram. But for a block inside a Super block, the definition of its symbolic parameters need not necessarily be done in the context of the Super block itself, it may be done in the context of the Super block containing the first Super block (if any), the Super block containing the Super block containing the first Super block (if any), etc., or the context of the top diagram. It can even be done in the Scilab environment. We will discuss this latter mechanism in Chapter 11.

The scoping rules that apply are fully consistent with Scilab's scoping rules: the definition of a variable is first searched in the local Super block context, if it is not found, the one-level up context is searched, and so on. This provides a rigorous methodology for using symbolic parameters avoiding any risk of name conflicts when copying a Super block from one location to another.

To understand the scoping rules, consider the example in Figure 7.19. The context of the main diagram defines the variables A, x, B and C. The context of the Super block defines B and y. Available parameters and their values inside this Super block are: A=3, B=4, x=%pi/7, y=1, C=1.

Only the symbolic parameters that are defined as Scilab variables in a context can be used as block parameter. The available symbolic parameters in any diagram can easily be inspected using the `Available parameters` button in the `View` menu; see Figure 7.20 for an example.

The scoping rules in Scicos also provides the proper mechanism for an intuitive masking operation. The masking operation consists of changing the behavior of a Super block in a diagram so that it looks just like a basic block. In particular, when we double click on a masked Super block, Scicos does not open a new editor window but opens up the standard block GUI used to set block parameters. When a masked Super block is created, these parameters, which are simply the parameters that are used inside the Super block but not defined inside the Super block, are identified automatically by Scicos. Let's consider the example in Figure 7.19 and suppose that the blocks inside the Super block use the variables x, y and B; see Figure 7.21.

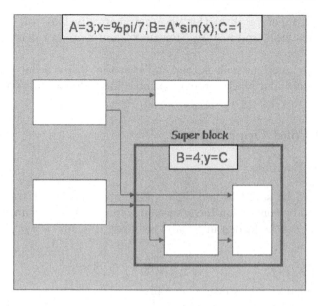

Figure 7.19. Inside the Super block `A=3`, `B=4`, `x=%pi/7`, `y=1`, `C=1`.

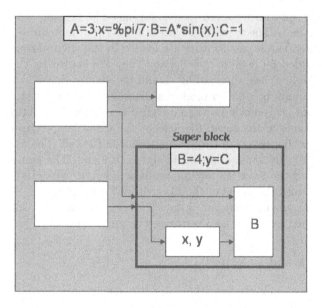

Figure 7.20. Available Scilab variables for use as block parameters.

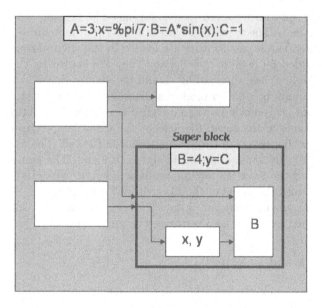

Figure 7.21. The parameters of the masked Super block will be `C` and `x`.

Masking of the Super block in this case automatically finds the parameters of the masked block, which are C and x. Why? Because C and x are not defined but are needed inside the Super block.

The use of Scilab functions such as load and exec that define Scilab variables implicitly should be avoided in Super block contexts for masking to function properly.

7.4 Save and Load Operations

7.4.1 Scicos File Formats

Scicos diagrams can be saved in two different formats. The most common format is a binary file for which by convention the extension .cos is used. This file is simply a binary Scilab data file and can be loaded in Scilab (assuming its name is myfile.cos) using the command

$\longmapsto$ load myfile.cos

There are three Scilab variables inside .cos files:

- scs_m: this Scilab structure contains all the information concerning the diagram
- %cpr: this Scilab structure contains the result of the compilation. It is an empty structure if the diagram had not been compiled before being saved (diagram compilation will be discussed later)
- scicos_ver: a string indicating the version of Scicos that has produced the diagram.

Normally, the user never needs to load this file in Scilab directly except for batch processing, which is discussed in Chapter 11. The loading is done in the Scicos environment using the Open button in the File menu. The file can also be loaded when launching Scicos as follows:

$\longmapsto$ scicos myfile.cos;

Scicos diagrams can also be saved in text format. The possible extensions of the text files are .cosf and .xml. The tt .cosf files are Scilab scripts that can be executed using the exec command, provided Scicos libraries are loaded in the environment. Note that Scicos libraries are not loaded by default into Scilab. They are loaded by the function scicos. So, just as in the binary case, the loading should be done by Scicos.

The expert user may be able to modify a diagram by editing the corresponding text file. But in most cases, it is much easier to use the Scicos graphical editor to do the editing.

Note that the name of the saved file, except the extension, corresponds to the name of the diagram. New diagrams are named by default Untitled. The name of the diagram can be changed using the Rename button in the File menu. The name is also changed if the file is saved, using the Save As button, under a different name. In that case, the name of the diagram becomes the name of the file (without the extension).

Finally note that when a diagram is saved window parameters, such as the position, size, zoom factor, etc., are saved and restored when it is opened.

7.4.2 Super Block and Palette

From within a Super block window, there are two saving options: one for saving the full diagram and another for saving the Super block alone. The contents of all the Super blocks present in a model are saved when the diagram is saved. So there is no need to save the content of a Super block unless we want to use it in the construction of another diagram. Super blocks can be saved just like any diagram in different formats. The saved file is identical to that of a main diagram. It can be loaded into a main Scicos diagram or inside a Super block.

A palette is also a standard Scicos diagram. Any Scicos diagram can be loaded as a palette. This means that blocks can be copied from it but the diagram itself cannot be edited. The button `Load as Palette` can be used to load any existing Scicos diagram as a palette. To avoid searching for palette files often used, Scicos provides the possibility of defining a list of palettes with their names and locations (see `Pal editor` in the menu `File`). These palettes can then be loaded simply by clicking on their names when the button `Palettes` is clicked. The default Scicos palettes are originally placed in this list.

7.5 Synchronism and Special Blocks

When two blocks are activated with the same activation source (for example the same `Event Clock`), we say that they are synchronized. In that case, they have the same activation times, and if the output of one is connected to the input of the other, the compiler makes sure the blocks are executed in the correct order. Synchronism is an important property. Two blocks activated by two independent clocks exactly at the same time are not synchronized. Even though their activations have identical timing, they can be activated in any order by the simulator.

On the other hand, two activations can be synchronous but not have exactly the same timings. For example, an activation can be a subset of another activation. Consider two activation sources, one generating a sequence of events with frequency 2 and another with frequency 1. In this case, half the events generated by the fast clock are simultaneous with the events of the slow clock. If these two activations are generated by two independent clocks, then they are not synchronized. So even when the two events are simultaneous, the blocks they activate can be activated in any order by the simulator. To enforce synchronism in this case, we have to make sure the two activations have the same source (same `Event Clock` for example). In this example this means that the slow activation is obtained from the fast activation.

There are two very special blocks in Scicos: the `If-then-else` and the event select blocks. They are in the `Branching` Palette. Even though they look like standard Scicos blocks and can be manipulated by the editor as such, strictly speaking they are not blocks. We shall give a detailed presentation of these blocks and the notion of synchronism in Chapter 8. Here, it suffices to say that these blocks are the only blocks that generate events synchronous with the incoming event that had activated them. These blocks, which can be considered as counterparts of conditional statements `If-then-else` and `Switch` in the C language, are used for conditional subsampling of activation signals.

Consider the diagram in Figure 7.22. The `If-then-else` block *redirects* the event it receives toward its `then` output port, activating the delay block `1/z` if its regular input is positive. If not, the activation goes out through the `else` port, which is not connected to anything. The `square wave generator` generates a series of alternating ones and minus ones. Every time the `1/z` block is activated, it adds the output of the generator to its content (initially set to zero). Since the activation takes place only when the output of the generator is positive, the content of `1/z` goes up by one at every other activation of the clock. This is confirmed by the result of the simulation illustrated in Figure 7.23. The period of the clock in this simulation is set to 1.

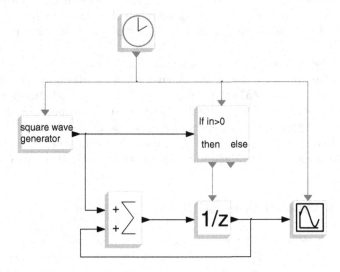

Figure 7.22. A synchronous Scicos diagram.

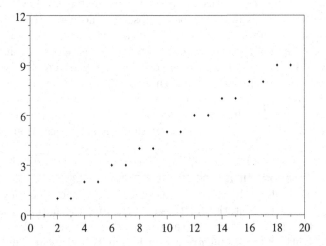

Figure 7.23. Simulation of Figure 7.22 showing that the counter goes up by one at every other activation of the clock.

There are other blocks in Scicos palettes that have special functionalities. For example, the GOTO and FROM blocks are used to establish connections without actually drawing links. These blocks are useful for connecting blocks that are situated in different Super blocks and at different levels of hierarchy.

The END block can be used to set the final time of simulation using a symbolic parameter defined in contexts. Note that the Scilab variables defined in the contexts can only be used for defining block symbolic parameters; they cannot be used for setting the simulation parameters, and in particular the final simulation time, in the Setup dialog of the Simulation menu.

8

Scicos Formalism

We have seen, in the previous chapter, how simple Scicos models can be constructed and simulated. To be able to construct more complex models, in particular models involving conditional and unconditional subsampling, it is important to understand the formalism on which Scicos is based. This is the subject of this chapter. The information in this chapter is also essential for reading the next chapter, which discusses the construction of new blocks.

8.1 Activation Signal

A simulation function is associated to each Scicos block. This function is called when the block is activated. The activation times for a given block are specified by the activation signals received on its activation input ports.

8.1.1 Block Activation

Memoryless Case

Consider the diagram in Figure 8.1. In this diagram, the activation signal (event) activates synchronously the three blocks Source, Func.1, and Func.2. This synchronous activation occurs because they are activated by the same activation signal. The order in which the associated functions are called is determined by the Scicos compiler. If we assume that the Func blocks are memoryless immediate functions, Func.1 needs the value of the output of the Source block before being called, and similarly Func.2 needs the output of Func.1, then the compiler orders the blocks accordingly.

In this example, the activation times of the three blocks were the same. However, having the same activation time does not necessarily imply synchronism. Consider the diagram in Figure 8.2. In this case the two event generators are identical and the events they generate have the same timing but are not synchronized. When a diagram contains more than one activation source, the compiler computes an order for each activation separately. During the simulation, it is the timing of the activations that determines the order in which the events are fired, and consequently the blocks are activated. If, as is the case here in Figure 8.2, two events have identical timing, then the order of firing is arbitrary. So Func.2 could be executed before or after the other two blocks. Depending on which one occurs, the result of the simulation can be completely different.

S.L. Campbell et al., *Modeling and Simulation in Scilab/Scicos with ScicosLab 4.4*,
DOI 10.1007/978-1-4419-5527-2_8, © Springer Science+Business Media, LLC 2010

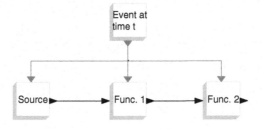

Figure 8.1. A synchronous diagram.

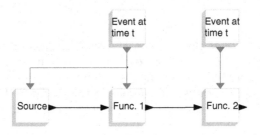

Figure 8.2. An asynchronous diagram.

Discrete Blocks with Internal State

If a block has an internal state, then when the block is activated, the simulation function corresponding to the block may be called more than once. In particular, if the block has an internal state and an output, then it will be called twice. Consider the diagram in Figure 8.3. The delay block is simply a memory block holding in its internal state the value of its input at the time of activation. Its output corresponds to the previous value of the state. At its activation, this block is called first with **flag** 1 so that it can compute its output (in this case a simple copy of the state), and then it is called with **flag** 2 to compute its state (copy of its input).

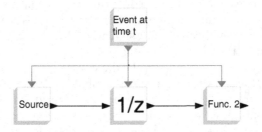

Figure 8.3. A synchronous diagram with a memory (delay) block.

The order in which block functions are called with **flag** 2, contrary to the case of **flag** 1, has no importance because the result has no bearing on the outputs of the blocks.

But in this case, even the order with which the blocks are called with `flag` 1 is not the same as in the case of the diagram in Figure 8.1. Here, the `1/z` block does not require the value of its input when it computes its output. This means that this block can very well be activated before the `Source` block. But in any case, `Func.2` comes after the `1/z` block.

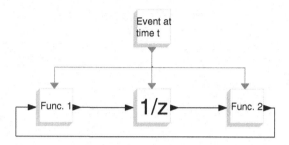

Figure 8.4. Diagram with feedback but without algebraic loop.

The blocks whose outputs do not depend directly on their inputs must be specified. Otherwise the compiler cannot compute the correct order of execution. In the example of Figure 8.3 even if `1/z` is not specified as such a block, the blocks can be ordered. However, in the example in Figure 8.4, the compiler cannot compute the order of block activations due to the presence of an algebraic loop. This is why the `dep_u` property of the `1/z` block is set to false and the compiler finds that the block functions must be called with `flag` 1 in the following order: `1/z` followed by `Func.2` followed by `Func.1`. And then `1/z` is called with `flag` 2.

General Case

A Scicos block can be more complex than just an internal state and an output. The event generator, for example, has an output activation port. We have seen in the previous chapter blocks with continuous-time internal states such as the integrator block. A general Scicos block can be fairly complex and will be discussed in the next chapter.

8.1.2 Activation Generation

The event generator in diagrams of Figures 8.1 through 8.4 generates an activation signal at a given time. This output activation is preprogrammed at the output activation port of this block. The block itself is never activated. Since only one activation can be preprogrammed at any output activation port, a Scicos block cannot serve as a clock event as we have seen in the previous chapter. In fact, the event clock is not a basic block; it is a Super block containing an event delay block with a feedback, as illustrated in Figure 8.5.

Initially an activation is programmed on the output activation port of the delay block. This activation, when fired, activates the delay block itself. Since this block has no internal state and no (regular) output, the corresponding function is not called with `flags` 1 and 2. Instead, it is called with `flag` 3 because it has an output activation port. In this case, the function schedules a new activation on its output activation port by returning the delay (in this case `T`), after which the scheduled activation must be fired. When this activation is fired, it schedules in turn a new activation, and so on. This way, an event clock with period `T` is constructed.

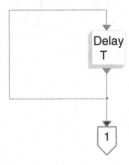

Figure 8.5. An event clock.

8.2 Inheritance

In the examples of the previous chapter, we had encountered many blocks without activation input ports. Consider the diagram in Figure 8.6. Normally the `Func.1` block should not be activated because it receives no activation signal. But by convention, in the absence of input activation ports, a block inherits its activation from its regular input.

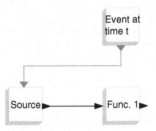

Figure 8.6. The block `Func.1` is activated by inheritance.

The inheritance mechanism is implemented at a precompilation phase. For example, in the case of the diagram in Figure 8.6, the precompiler adds the missing activation port and link as illustrated in Figure 8.7. The user never sees this diagram; the precompilation phase is completely transparent to the user.

If the block has more than one regular input, the inheritance mechanism places as many activation input ports on the block as the block has regular input ports. See the diagram in Figure 8.8 and the diagram obtained after the precompilation phase illustrated in Figure 8.9. The + block outputs an activation signal with activation times corresponding to the union of input activation times.

We see here that a block can have more than one activation input port. In this case, the block is activated when it receives an event on one or the other of its input activation ports. Thus the block is activated at the union of the two activation times. This explains how `Func.2` inherits its activation.

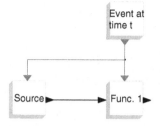

Figure 8.7. The diagram after the precompilation phase.

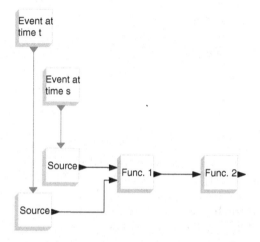

Figure 8.8. Blocks `Func.1` and `Func.2` are activated by inheritance.

8.3 Always Active Blocks

There is another way a block can be active, and that is by being declared *always active*. Consider the sinusoid generator in the diagram of Figure 7.4 in Chapter 7. This block is not explicitly activated, and it clearly does not inherit. But it is active because it is declared always active. This is done by setting the block's `dep_t` property to true.

An always active block is, at least as far as the formalism is concerned, always active. But during the simulation, the block is activated only when needed. For example, in the diagram of Figure 7.4, the output of the sinusoid generator should evolve continuously. But since this value is used only by the scope at its activation times, during the simulation, the function associated with the sinusoid generator is only called at these times.

In the diagram of Figure 7.6, the integrator block is also declared always active. But even if it were not, it would not have made any difference because it would have inherited *always activation* from the sinusoid generator. In this case, since the output of the sinusoid generator affects a differential equation, the times that the function associated with the sinusoid generator is called, in addition to the activation times of the scope, are the times (integration time steps) imposed by the numerical solver.

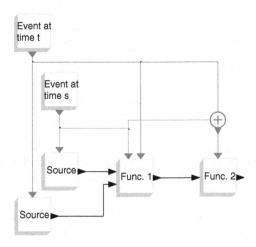

Figure 8.9. The diagram after the precompilation phase.

8.4 Constant Blocks

A block without input, not explicitly activated and not being always active, is a constant block. A constant block's outputs do not evolve as a function of time. During the simulation, the function associated with a constant block is called only at the initialization phase. The same holds for blocks inheriting from constant blocks. See, for example, the diagram in Figure 8.10. If Source is not always active, then it is a constant. If Func.1 is not always active, it inherits its activation, or more specifically its absence of activation, from the Source block. The Func.1 block's function is called just once to compute its output after the initialization phase. This is, of course, reasonable because the output of Func.1 does not evolve.

Figure 8.10. Inheritance from a constant block.

8.5 Conditional Blocks

In many applications, in particular in signal processing, signals with different frequencies interact in the same model. A common situation is the decimation operation, which consists in generating a signal B from a signal A by taking one out of every n values of A. The frequency of signal B is then that of signal A divided by n. To implement this operation, one might consider using two independent clocks to fix the two frequencies. This is done in the diagram of Figure 8.11 using a sample and hold block.

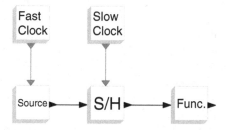

Figure 8.11. Incorrect way of implementing decimation.

But we have seen that two activation sources generate asynchronous events. So even if we set the period of the slow clock to exactly n times that of the fast clock, at times when the two events have the same time, the order of block execution is not predictable.

We see through this example that there is a need for being able to define synchronism between two nonidentical activation signals. The events generated by the fast clock do not always have the same timing information as that of the slow clock. However, when they do (one out of every n times), they must be considered synchronized. This is particularly important when the fast and slow signals are used in common operations later such as being added together.

This type of synchronous signal can be constructed in Scicos using two special blocks: the **If-then-else** and the event select, which can be found in the **Branching** palette. Strictly speaking, these blocks are not Scicos blocks. They do not generate new activation signals. Rather they redirect the activation they receive.

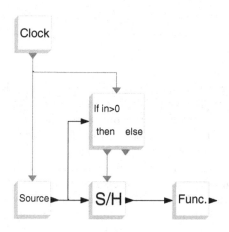

Figure 8.12. Conditional subsampling.

Consider the diagram in Figure 8.12. The sample and hold block, when activated, copies its input on its output. The **If-then-else** block activates the sample and hold block when the value of its input is positive. The output of this block is then a signal with an activation that depends on values of the signal generated by the **Source** block.

We say then that this signal is obtained by conditional subsampling. In this case, since the Source and S/H have identical activation sources (remember that the If-then-else block only redirects activations), they are considered synchronous by the Scicos compiler and ordered properly. In particular, in this case we can be sure that the output of the S/H block is always positive.

Now we return to the decimation problem previously discussed. The diagram in Figure 8.11 did not function properly because the Source and the S/H blocks were not synchronized. Using the If-then-else block, we can implement the decimation operation correctly as illustrated in Figure 8.13. The Counter Modulo block counts up to $n-1$ and then returns to zero. The S/H block is activated only when the output of the counter block is zero (activation through the else branch), that is, one out of every n times. And in this case, the Source and S/H blocks are synchronized.

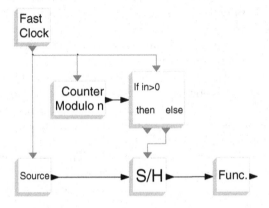

Figure 8.13. A correct way to implement decimation. The (Super) block freq_div can replace the Counter Modulo and If-then-else blocks.

The combination Counter Modulo and If-then-else provides a versatile mechanism for implementing frequency division. The division factor can be set by fixing the value of n, and the phase by the initial state of the counter. In fact, that is the way the freq_div (Super) block, available in the Events palette, is constructed.

The above process for generating synchronous clocks with different frequencies can become cumbersome for the designer when one frequency is not a multiple of another and/or when there are more than two frequencies involved. That is why Scicos provides a special block called SampleCLK. The SampleCLK block is used very much like a regular Event Clock however the activation signals generated by such blocks anywhere in a model are synchronized. For these blocks, Scicos finds automatically the smallest frequency from which the frequency of SampleCLK blocks can be derive and implements all the corresponding conditional blocks. See Figure 8.14. In this case, the slowest clock generating synchronous clocks of periods 3 and 5 by subsampling is a clock with period 1.

Conditional blocks work also for "always active" activation. Consider the diagram in Figure 8.15. The If-then-else block is always active (inheriting its activation from the sine generator) but activates the S/H block only when its input is positive. So the integrator receives on its input $\sin(t)$ if it is positive, and zero otherwise. Zero is the last value output

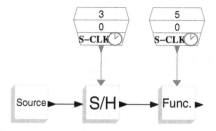

Figure 8.14. `SampleCLK` blocks generate synchronous events.

by S/H before deactivation. When a block is not active, its output remains constant. The simulation result is given in Figure 8.16.

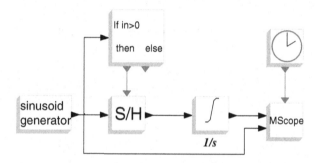

Figure 8.15. Conditional blocks function also with continuous-time activation.

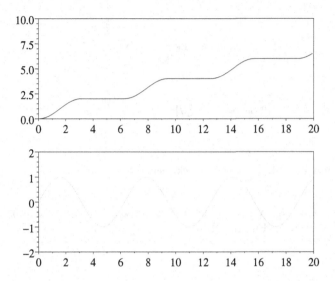

Figure 8.16. Simulation result. At the bottom we have the output of the **sinusoid generator** $\sin(t)$ and on top $\int_0^t \max(\sin(\tau), 0)d\tau$.

9

Scicos Blocks

A block is a basic module from which Scicos diagrams are constructed. A block corresponds to an operation, and by interconnecting blocks through links, we construct (implement) an algorithm. So in a sense, blocks in Scicos are the counterpart of functions in a programming language, and as in programming languages, some blocks are provided through standard libraries, while others are developed by the user.

Most Scicos users never need to develop a new block because standard libraries contain a large number of blocks providing most of the elementary functionalities needed in the construction of Scicos diagrams; moreover, blocks such as `Mathematical Expression` allow the user to define block functionality in terms of Scilab expressions without loss of efficiency (the Scilab expression in this case is compiled). That is why readers can skip this chapter on a first reading and come back to it when they want to learn about block construction.

9.1 Block Behavior

Scicos blocks can be connected through input and output ports, and, activation input and output ports. Input and output ports are used to communicate signals, which in general can be matrices of different types. In Scicos 4.4, the following types are supported: double, complex double, int8, uint8, int16, uint16, int32, uint32 and Boolean. Activation ports transmit activation signals.

A Scicos block can be a complex entity. It can have multiple inputs and outputs, a continuous-time vector state, a list of discrete-time matrix states, zero-crossing surfaces, etc., as shown in Figure 9.1. But most Scicos blocks do not contain all of these features simultaneously.

The behavior of a block is primarily specified by the way it is activated. We have seen in the previous section that a block is activated in three different ways. The way the block learns about the way it is activated is via an activation index `nevprt`.

9.1.1 External Activation

A block is activated when it receives activations on its activation input ports. As we have seen in the previous section, such activations can also be inherited when the block has no activation input ports, in which case the corresponding activation ports are added by the Scicos compiler and the activations are considered to be external.

S.L. Campbell et al., *Modeling and Simulation in Scilab/Scicos with ScicosLab 4.4*,
DOI 10.1007/978-1-4419-5527-2_9, © Springer Science+Business Media, LLC 2010

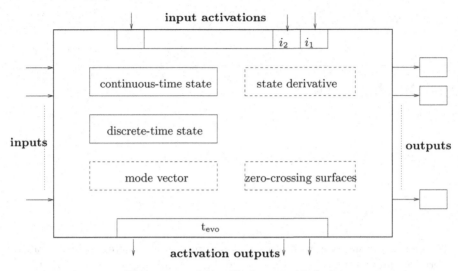

Figure 9.1. Scicos block: an inside look.

In the case of external activation, the activation index **nevprt** is a positive integer obtained by the following formula:

$$n_{\text{evprt}} = \sum_{j=1}^{n} i_j 2^{j-1}, \qquad (9.1)$$

where n is the number of input activation ports and i_j is equal to one or zero depending on whether an activation has been received on port j. Note that **nevprt** corresponds to the binary coded number $i_n i_{n-1} \cdots i_1$.

In some cases, block behavior does not depend on **nevprt**. For example, the summation block performs an addition regardless of the value of **nevprt**. But the **Selector** block uses **nevprt** to select the input that should be copied into the output.

Event Activation

When a block is activated in this way by an event, say at time t_e, it can update its outputs:

$$y(t_e) = f_1(t_e, x(t_e^-), z(t_e^-), u(t_e), \mu(t_e)). \qquad (9.2)$$

In the above, t_e denotes the event time, and $x(t_e^-)$ and $z(t_e^-)$ denote respectively the continuous-time and discrete-time states just before the occurrence of the event. $y(t_e)$ denotes the outputs of the blocks (which can be more than one and each a vector or matrix of arbitrary size and data type). $u(t_e)$ denotes similarly the inputs.

The function f_1 depends also on n_{evprt}, which is part of μ:

$$\mu = (n_{\text{evprt}}, m, p).$$

Here m and p represent respectively the **mode** and the **simulation phase**. We shall discuss them later.

If the block contains output activation ports, upon activation it can program events on them by providing the delay time to event firing on each activation output port:

$$t_{\text{evo}} = f_3(t_e, x(t_e^-), z(t_e^-), u(t_e), \mu(t_e)). \qquad (9.3)$$

Finally, if it has internal states, it can update them (both continuous-time and discrete-time):

$$[z(t_e), x(t_e)] = f_2(t_e, x(t_e^-), z(t_e^-), u(t_e), \mu(t_e)). \tag{9.4}$$

The vector t_{evo} is generated at block activation (if the block has any output activation ports). If the jth entry is positive, then an event is programmed on the jth output activation port of the block. The activation date of this event is obtained by adding t_e to this entry.

Continuous-Time Activation

If a source of block activation is "always activation," then activation occurs over time intervals and not at specific times as in the case of event activation. See, for example, the diagram in Figure 8.15. In this case, the output calculation given in equation (9.2), thanks to the continuity assumption of all the signals over a continuous-time activation period, can be simplified as follows:

$$y(t) = f_1(t, x(t), z(t), u(t), \mu(t)). \tag{9.5}$$

The output activation programming and state updates do not apply in this case, but if the block has a continuous-time state, it evolves according to the following differential equation,

$$\dot{x} = f_0(t, x(t), z(t), u(t), \mu(t)) \tag{9.6}$$

over the activation period, which we call the **simulation phase** 2, as opposed to event activation instances which are referred to as **simulation phase** 1 . Note that $z(t)$, like n_{evprt}, is constant over this period. So is the mode m if there is one.

Mode and Zero-Crossing

The mode parameter is not part of the Scicos formalism. It is introduced to facilitate the implementation of the numerical solver. Most blocks that a user may have to construct don't need any mode or zero-crossing, and the user can use existing blocks with mode and zero-crossing without a complete understanding of the way they work. For the sake of completeness, however, we give a short presentation below. More details can be found in the examples provided later in this chapter.

Numerical integrators can experience step size control and other problems at places where the functions being integrated are not continuously differentiable. If f_1 or f_0 is not smooth (continuously differentiable) at some points, then mode is used in such a way as to make sure the numerical solver never encounters these points of discontinuity inside an integration interval.

Consider the following simple example: $y = u$ if $u \geq 0$, and $y = -u$ otherwise. This function realizes the absolute-value function and is not differentiable at 0. In this case, a mode can be defined to specify at the start of the integration period whether u is positive. To make sure the integration stops when the sign of u changes, a zero-crossing surface is introduced at zero. During the integration period (which could end because of the zero-crossing), the output y is computed as follows: $y = u$ if $m = 1$, and $y = -u$ otherwise. After the zero-crossing the mode is recomputed and the integration continues. The computation of the mode and the zero-crossing surface are performed by the block. If the **simulation phase** is 1, we have

$$[m(t), s(t)] = f_9(t, x(t), z(t), u(t), \mu(t)). \tag{9.7}$$

The zero-crossing surfaces $s(t)$ may be present even if the block has no mode. When a zero-crossing occurs, the block is internally activated, which means that (9.3) and (9.4) are called with $n_{evprt} = -1$. The blocks in the Threshold palette, for example, have zero-crossings but no modes.

9.1.2 Always Activation

A block can be declared always active. This is, for example, the case of the sine generator block. In this case the nevprt index is zero. Always activation can be thought of as an activation received on a fictitious activation input port 0 that would correspond to $n_{evprt} = 0$.

Always activation is a special case of continuous-time activation where the activation period is equal to the period of simulation. So the behavior of the block is specified again by equations (9.5) and (9.6) with n_{evprt} set to zero.

9.1.3 Internal Zero-Crossing

When a zero-crossing occurs inside a block, this block is activated with n_{evprt} equal to -1 at the time of crossing. If this time is denoted by t_e, then the state update formula is given by (9.4) with $n_{evprt} = -1$. This type of activation is used, for example, in the bouncing balls demo block of Scicos to change the direction of balls (characterized by the continuous-time state of the block) after a collision.

9.2 Blocks Inside Palettes

Scicos libraries of blocks are called palettes. There are a number of standard palettes available in Scicos providing many elementary blocks.

The available palettes are the following:

- **Sources library.** It includes signal generators such as the square-wave, sawtooth, and sine-wave generators, and also read from file blocks. The read from file blocks generate signals from data stored on files.
- **Sinks library.** It includes blocks that store signals on a file or display signals on a screen. These blocks have no output ports.
- **Branching library.** This palette contains switches, multiplexers, selectors, and a relay. It also includes two very special blocks that redirect activation signals: the If-then-else and the events select blocks.
- **Non_linear library.** The blocks in this palette realize memoryless nonlinear operators such as trigonometric functions, multiplication, logarithm, inversion, and saturation. The palette also includes interpolation and table lookup blocks, and a special block called Mathematical Expression used to express a nonlinear operator in terms of a Scilab expression. The constraint on the scilab expressions is that only scalar expressions can be used, but the evaluation is very efficient.
- **Events library.** This library includes blocks for generating and manipulating events (activation signals). It also includes the block Stop, which halts the simulation if activated.
- **Threshold library.** The blocks included in this palette can generate events when a signal crosses zero.

- **Others library.** This palette includes a number of blocks useful for the construction of new blocks: `Cblock4`, `Scifunc`, `GENERIC`, ... These blocks are discussed later in this chapter. It also includes logical and relational blocks. It also includes the `Text` block, which is not a real block, but a way to place text inside a diagram.
- **Linear library.** This palette contains blocks realizing linear operators such as summation, gain, and many dynamical linear systems in both continuous and discrete time. Delay blocks and jump linear systems are also included in this palette.
- **Matrix library.** This palette includes elementary matrix operations. In Scicos 4.4, inputs and outputs can be matrices.
- **Integer library.** Bitwise and logical operations requiring integer data types are provided in this palette.
- **Lookup Tables library.** Various interpolation routines are available as Scicos blocks in this palette.
- **Modelica library.** All Modelica blocks are included in this hierarchical palette that includes an `electrical` palette and a `Thermo-Hydraulic` palette.

The above palettes provide general-purpose blocks. Specialized palettes can be added by the user.

9.3 Modifying Block Parameters

Most blocks have parameters that must be changed by the user for specific applications. Block parameters can be changed by "opening" a block (clicking at the center of the block). This action opens a dialog box, which can be used to define new parameter values.

Blocks parameters are stored in two different ways:

1. Symbolically: the exact expression entered for each parameter in the dialog box is stored as Scilab text. For example, the user can enter `2*%pi`, and this expression is saved and shown the next time the dialog box is opened.
2. Numerically: the expression is evaluated and the numerical value is saved.

It may seem redundant to store parameter values in two different ways but there are good reasons for doing this. One reason is that it is more convenient to see `2*%pi` the next time the dialog box is opened than it is to see `6.2831853`. But the main reason is that the values of the parameters can be defined in terms of Scilab variables already defined in the context of the diagram. For example, a parameter can be set to `sin(x)*y`. The numerical value of this parameter is evaluated based on the values of x and y defined in the context of the diagram. Keeping the symbolic expression allows Scicos to update the value of the parameter when x or y is modified in the context. In this case, x and y are symbolic parameters.

9.4 Super Block and `Scifunc`

In addition to the blocks available in Scicos palettes, the user can create and use custom blocks by adapting existing blocks. There are various ways of doing this.

- using the Super block facility,
- using `Scifunc` blocks, which allow block functionality to be defined on-line by Scilab expressions.
- using C or Scilab programs, dynamically linked with, or loaded in, Scilab.

The third method, which will be presented in the next section, gives the best results as far as simulation performance is concerned, provided a C program is used.

9.4.1 Super Blocks

To construct a Super block, the user should copy the "Super block" block from the Others palette into the Scicos window and click on it. This will open up a new Scicos window in which the Super block should be defined. Super blocks can also be defined using the Region-to-Super-block facility, which creates a Super block from existing blocks in a specified region of the diagram.

The construction of the Super block model is done in the same way as the earlier diagrams, that is, by copying and pasting blocks from various palettes into the Super block's editor window and connecting them. Input and output ports of the Super block should be specified by input and output block ports available in the Sources and Sinks palettes. Super blocks can, of course, be used within Super blocks.

A Super block can be saved in various formats. If the Super block is of interest in other constructions, it can be converted into a block using the Masking operation discussed in Chapter 7 and placed in a user palette. This can be done by the SaveBlockGUIc facility.

Once the Super block, or more specifically the corresponding Scilab interfacing function, is saved, it can be used in Scicos. This can be done using Add-new-block in the Edit menu. The block can also be placed in a palette for future use. See Section 9.6.

Under certain conditions, the dynamics of the content of a Super block can be expressed using a single C program. Code-Generation generates this program automatically; see Section 12.

If the Super block is used only in a particular construction for one Scicos diagram, then the Super block need not be saved. A click on the Exit will close the Super block window and activate the main Scicos window (or another Super block window if the Super block was being defined within another Super block). Saving the block diagram automatically saves the content of all Super blocks used inside of it.

A Super block, like any other block, can be duplicated using the Copy button. In that case all the contents of the Super block are duplicated.

9.4.2 Scifunc

The Scifunc block enables using the Scilab language for defining, on-line, new Scicos blocks. Block information, such as the number of inputs and outputs, initial states, the initial firing vector, and information regarding the behavior of the block during simulation have to be entered by the user. This input consists essentially of Scilab expressions defining functions f_i (presented in Section 9.1). This is done interactively by clicking on the Scifunc block once it is copied into the Scicos window. The main disadvantage of Scifunc is that the dialog for updating block parameters cannot be customized. In order to have a customized dialog box, a Scilab function called the interfacing function must be developed. We shall see how this is done in Section 9.5.1.

It should be noted that in the case of Scifunc, like all older types of Scicos blocks, the function f_3 must return the time of the outgoing events and not the delay with respect to the activation time of the block.

9.5 Constructing New Basic Blocks

In this section we show how a new basic block can be constructed and used in Scicos.

Each Scicos basic block is defined by two functions. The first one, which must be in the Scilab language, handles the interactions with the editor. It is this function that specifies what the geometry of the block should be, how many inputs and outputs it should have, what the block type is, etc. It is also this function that handles the user interface (updating block parameters and initializing the states). This function is referred to as the *interfacing function*.

The second function, which is called the *computational function*, is normally written in C, but can also be a Scilab function. This function defines the behavior of the block during the simulation. For simple blocks, this amounts to coding the f_i functions which have to be evaluated by the program depending on the value of an input flag. The i in f_i corresponds to the value of the flag. We shall come back to this in Section 9.5.2. In order to get the best performance, this part of the Scicos block should be defined by C programs, but Scilab functions are often used in the early phases of development for testing and debugging purposes since they are easier to write.

Even though each basic block requires both an interfacing function and a computational function, the user often needs to be developed only one. For example, a computational function can be used with the GENERIC block available in the Others palette. The GENERIC block is a generic interfacing function. It suffices to specify the "name of the computational function" in the dialog box of the GENERIC block along with information concerning the input/output numbers and sizes, initial states, etc. The name of the program, in the case of C programs, is the name of the entry point when the C program is incrementally linked with Scilab (see Section 9.5.2). In the case of a Scilab program, it is simply the name of the Scilab function. Scilab functions must, of course, be loaded as usual into Scilab before Scicos is launched. This can be done using either getf or load if the function is contained in a library. See Section 2.5.

The drawback of using the GENERIC block, as in the case of Scifunc, is the generic dialog box. In most cases, it is desirable to customize both the block's dialog box and its look. For that, a custom interfacing function must be used. This Scilab function should provide information on the geometry of the block and handle the dialog for defining and updating block parameters and states. The interfacing function is used by the graphical editor, as opposed to the computational function, which is used by the simulator. Section 9.5.1 describes how the interfacing function and the underlying programs should be developed.

On the other hand, in some cases a new block may consist simply of a new interfacing function using an existing computational function. Consider, for example, the two continuous-time *linear dynamical system* blocks in the Linear palette. One is in state-space form and the dialog requests the A, B, C, D matrices. The other uses transfer functions and the user should enter the transfer function of the system. But both of these blocks use the same computational function, namely csslti. The reason is that the transfer function is converted to a state-space representation in the corresponding interfacing function. So two blocks may look very different but have a common computational function.

A new block realizing a low-pass filter should have a dialog where information such as the cut-off frequency, order, overshoot, etc., are specified. But once the filter is designed in the interfacing function (using specialized Scilab functions), the simulation can very well be performed with csslti. The reason is that the filter is nothing but a linear system characterized by A, B, C, D matrices.

9.5.1 Interfacing Function

The interfacing function is used only by the Scicos editor. In particular, it is used to initialize, draw, connect the block, and to modify its parameters. In addition, the interfacing function initializes block states and specifies certain block properties. It also specifies the name of the computational function to be used by the simulator. What the interfacing function should return depends on an input flag job.

Syntax

```
[x,y,typ]=block(job,arg1,arg2)
```

Parameter job

The parameter job can take on the following values:

- 'plot': the function draws the block and its label. arg1 is the data structure of the block. arg2,x,y,typ are unused.
  In general, the standard_draw function can be used. It draws a rectangular block.
- 'getinputs': the function returns the position and type of input ports (regular and event). arg1 is the data structure of the block. arg2 is unused. x is the vector of x coordinates of input ports. y is the vector of y coordinates of input ports. typ is the vector of input ports types.
  In general, the standard_input function can be used.
- 'getoutputs': returns position and type of output ports (regular and event). arg1 is the data structure of the block. arg2 is unused. x is the vector of x coordinates of output ports. y is the vector of y coordinates of output ports. typ is the vector of output ports types .
  In general, the standard_output function can be used. The standard_input and standard_output functions place regular input/outputs on the side of the block and the activation inputs on top and activation outputs at the bottom of the block.
- 'getorigin': returns coordinates of the lower left point of the rectangle containing the block's shape. arg1 is the data structure of the block. arg2 is unused. x is the vector of x coordinates of output ports. y is the vector of y coordinates of output ports. typ is unused.
  In general, the standard_origin function can be used.
- 'set': This function opens a dialog for block parameter acquisition. arg1 is the data structure of the block. arg2 is unused. x is the new data structure of the block. y is unused. typ is unused.
- 'define': provides initialization of the block's data structure. It sets block's initial type, number of inputs, number of outputs, etc. arg1, arg2 are unused. x is the data structure of the block. y is unused. typ is unused.

Example: the Interfacing Function of the Absolute Value Block

```
function [x,y,typ]=ABS_VALUE(job,arg1,arg2)
  x=[];y=[];typ=[]
  select job
    case 'plot' then
    standard_draw(arg1)
    case 'getinputs' then
    [x,y,typ]=standard_inputs(arg1)
    case 'getoutputs' then
    [x,y,typ]=standard_outputs(arg1)
    case 'getorigin' then
```

```
  [x,y]=standard_origin(arg1)
  case 'set'  then
  x=arg1;
  graphics=arg1.graphics;exprs=graphics.exprs
  model=arg1.model;
   while %t do
    [ok,zcr,exprs]=..
        getvalue('Set block parameters',..
                 ['use zero_crossing (1: yes) (0:no)'],..
                 list('vec',1),exprs)
     if ~ok  then  break, end
    graphics.exprs=exprs
     if ok  then
       if zcr<>0  then
        model.nmode=-1;model.nzcross=-1;
       else
        model.nmode=0;model.nzcross=0;
       end
       x.graphics=graphics;x.model=model
       break
     end
   end
  case 'define'  then
  nu=-1
  model=scicos_model()
  model.sim=list('absolute_value',4)
  model.in=nu
  model.out=nu
  model.nzcross=nu
  model.nmode=nu
  model.blocktype='c'
  model.dep_ut=[%t %f]
  exprs=[string([1])]
  gr_i=['txt=[''ABS''];';
        'xstringb(orig(1),orig(2),txt,sz(1),sz(2),''fill'')']
  x=standard_define([2 2],model,exprs,gr_i)
  end
endfunction
```

The data structures used in the interfacing function are explained in Appendix A. The only hard part of defining an interfacing function is in the 'set' case. Note that the function getvalue is used to create a dialog for the block. This should be done in all user-defined blocks as well, since this function is overloaded, when Eval is performed, to read and evaluate block parameters. getvalue is a generic dialog box with built-in type checking facilitating the implementation of Scicos blocks' dialog boxes.

9.5.2 Computational Function

The computational function is called in various ways by the simulator. The way the block is called (the calling sequence) is characterized by the *type* of the interfacing function. It is strongly recommended that type 4 and 5 computational functions now be used. The other types are now obsolete. Type 4 is used for programs written in C and type 5 for Scilab computational functions.

Type 4 C functions receive two arguments: a structure containing block information and a flag.

flag

The flag indicates the job that the computational function must perform. It usually consists in updating some of the fields of the block structure. Specifically, the following jobs exist:

- **Initialization:** If the function is called with flag=4, then the continuous and discrete states can be initialized, or more specifically, reinitialized (if necessary) because they are already initialized by the interfacing function. The block output can also be initialized. This option is not used in most blocks. It is used by the ones that read and write data from files, for opening the file, or by scope for initializing the graphics window. It is also used by blocks, which require dynamically allocated memory; the allocation is done under this flag. Each block is called once and only once at the beginning of the simulation with flag 4.
- **Output update:** When flag=1, the simulator is requesting the outputs of the block. The computational function should use the needed information within the block's structure (inputs, states, nevprt, which codes the way by which the activation has arrived, etc.) to compute the outputs and place them at the addresses given in the block structure. Note that if the block contains different modes, then the way the output is computed depends on the *simulation phase*. The simulation phase indicates whether the call is made in the numerical integration phase or is due to an event.
- **State update:** When the simulator calls the function with flag=2, it means that an event has activated the block (nevprt$\neq$ 0). The activation may also be due to an internal zero-crossing event, in which case nevprt is equal to -1. With flag= 2, the simulator is asking the block to update its states x and z. Scicos provides the addresses of the states, where old values of x and z reside, and the block should update them in place. This avoids useless and time-consuming copies, in particular when part or all of x or z is not to be changed. If a temporary workspace is needed to perform the computations, it can either be statically allocated in the C routine or dynamically allocated when the block is called with flag 4.
  If nevprt$= -1$, indicating that the activation is due to an internal zero-crossing, then the vector jroot specifies which surfaces have been crossed and in which direction. In particular, if the ith entry of jroot is 0, then the ith surface (counting from zero) has not crossed zero. If it is $+1$ (respectively -1), then it has crossed zero with a positive (respectively negative) slope.
- **Integrator calls:** During the integration, the solver calls the function for the value of $\dot{x}$. This is done with flag=0. The computational function must compute $\dot{x}$ and place it in the address provided in the block structure.
- **Mode and zero-crossing:** Scicos solvers assume model smoothness. This means that the behavior of blocks, which contain or affect continuous-time states of the system must be smooth, or at least piecewise smooth. In this latter case, the points of nonsmoothness must be specified in such a way that the solver can, if needed, issue a cold restart when going through such points. An example of such a block is the ABS block. The absolute value function is smooth except at zero, where it is not even differentiable. In this case, we say that the absolute value block has two **modes**: one where the output equals the input, and one where the output is the negative of the input. A block may have many different modes, which are stored in the block structure. The point of nonsmoothness is represented by a zero-crossing, and the mode is updated after such

a crossing occurs. flag= 9 is used both for evaluating the zero-crossing function and setting the mode. Mode setting is done only in the simulation phase 1.

Note that the block may have more zero-crossings than modes. For example, the zcross block, which generates an event when its input crosses zero, has one zero-crossing surface but no mode.

- **Event scheduler:** To update the event scheduler table, the simulator calls the function with flag=3. This happens if the block has output activation ports. In this case the block has access to jroot.

- **Ending:** Once the simulation is done or at user request (by responding End in the Run menu after a Stop), the simulator calls each computational function with flag=5, once. This is useful, for example, for closing files that have been opened by the block at the beginning or during the simulation, to flush buffered data, and to free allocated memory.

Table 9.1 summarizes the role of each flag. Note that in most cases the computational functions are not called with all these flag values. For example, if a block has no output port, its associated computational function is never called with flag 1. If it has no continuous-time state, then it is not called with flag 0, etc. Also note that in this table we have not specifically stated as input time-invariant quantities such as the parameters rpar and ipar. The function clearly has access to these parameters and can use them in computing its outputs. Finally, the workspace is not mentioned in the table because it can be used in various ways. If needed, the workspace w is in general allocated when flag= 4 and freed when flag= 5. The workspace can be used as storage and in many different ways.

flag	inputs	outputs	description
0	t, nevprt, x, z, inptr, mode, phase	xd	compute the derivative of continuous time state
1	t, nevprt, x, z, inptr, mode, phase	outptr	compute the outputs of the block
2	t, nevprt>0, x, z, inptr	x, z	update states due to external activation
2	t, nevprt=-1, x, z, inptr, jroot	x, z	update states due to internal zero-crossing
3	t, x, z, inptr, jroot	evout	program activation output delay times
4	t, x, z	x, z, outptr	initialize states and other initializations
5	x, z, inptr	x, z, outptr	final call to block for ending the simulation
6	x, z, inptr	x, z, outptr	re-initialization of outputs and states
7			only used for internally implicit blocks
9	t, phase=1, nevprt, x, z, inptr	g, mode	compute zero-crossing surfaces and set modes
9	t, phase=2, nevprt, x, z, inptr	g	compute zero-crossing surfaces

Table 9.1. This tables illustrates the jobs that the computational function must perform for different flags.

The computational function receives most of its inputs from the `block` structure. A number of utility C function are available to allow access to the `block` structure. It is recommended to use these macros and not access directly the fields of the structure to assure compatibility with future versions of Scicos. A detailed description of available macros is provided below.

Inputs/outputs

These macros give access to block structure fields associated with regular inputs and outputs. They provide information about the number of inputs and outputs, their sizes and types. They also provide pointers to the corresponding data for reading and writing. See Table 9.5.2 for the list of available macros and a short description of each one.

Activations ports

The macros needed to process block activation information are described in Table 9.5.2.

Parameters

The C macros giving access to various block parameters are given in Table 9.5.2.

States and the workspace

A Scicos block can have continuous-time and discrete-time states. The block can also create additional work space by using memory allocation. The macros in Table 9.5.2 can be used to access the fields associated with states and workspaces.

Zero-crossing surfaces and modes

The zero-crossing and mode selection are advanced features that are used in Scicos blocks with hybrid dynamics. The associated macros are given in Table 9.5.2

Macro	Type	Description
GetNin(blk)	int	Returns the number of regular input ports.
GetInPortRows(blk,x)	int	Returns the number of rows (first dimension) of the regular input port numbered x.
GetInPortCols(blk,x)	int	Returns the number of columns (second dimension) of the regular input port numbered x.
GetInPortSize(blk,x,y)	int	Returns the regular input port size numbered x (y=1 for the first dimension, y=2 for the second dimension).
GetInType(blk,x)	int	Returns the type of the regular input port numbered x.
GetInPortPtrs(blk,x)	void *	Returns the regular input port pointer of the port number x.
GetRealInPortPtrs(blk,x)	double *	Returns the pointer of real part of the regular input port number x.
GetImagInPortPtrs(blk,x)	double *	returns a pointer to the imaginary part of the regular input port number x.
Getint8InPortPtrs(blk,x)	char *	returns a pointer to the int8 typed regular input port number x.
Getint16InPortPtrs(blk,x)	short *	returns a pointer to the int16 typed regular input port number x.
Getint32InPortPtrs(blk,x)	long *	returns a pointer to the int32 typed regular input port number x.
Getuint8InPortPtrs(blk,x)	unsigned char*	returns pointer to the uint8 typed regular input port number x.
Getuint16InPortPtrs(blk,x)	unsigned short *	returns pointer to the uint16 typed regular input port number x.
Getuint32InPortPtrs(blk,x)	unsigned long *	returns a pointer to the uint32 typed regular input port number x.
GetSizeOfIn(blk,x)	int	Returns the size of the type of the regular input port number x in bytes.
GetNout(blk)	int	Returns the number of regular output ports.
GetOutPortRows(blk,x)	int	returns number of rows (first dimension) of the regular output port number x.
GetOutPortCols(blk,x)	int	Returns the number of columns (second dimension) of the regular output port number x.
GetOutPortSize(blk,x,y)	int	Returns the size of the regular output port number x. (y=1 for the first dimension, y=2 for the second dimension)
GetOutType(blk,x)	int	Returns the type of the regular output port number x.
GetOutPortPtrs(blk,x)	void *	Returns a pointer to the regular output port number x.
GetRealOutPortPtrs(blk,x)	double *	Returns a pointer to the real part of the regular output port number x.
GetImagOutPortPtrs(blk,x)	double *	Returns a pointer to the imaginary part of the regular output port number x.
Getint8OutPortPtrs(blk,x)	char *	Returns a pointer to the int8 typed regular output port number x.
Getint16OutPortPtrs(blk,x)	short *	Returns a pointer to the int16 typed regular output port number x.
Getint32OutPortPtrs(blk,x)	long *	Returns a pointer to the int32 typed regular output port number x.
Getuint8OutPortPtrs(blk,x)	unsigned char *	Returns a pointer to the uint8 typed regular output port number x.
Getuint16OutPortPtrs(blk,x)	unsigned short *	Returns a pointer to the uint16 typed regular output port number x.
Getuint32OutPortPtrs(blk,x)	unsigned long *	Returns a pointer to the uint32 typed regular output port number x.
GetSizeOfOut(blk,x)	int	Returns the size of the type of the regular output port number x in bytes.

Table 9.2. Input/output C macros

Macro	Type	Description
GetNevIn(blk)	int	Returns nevprt, the input activation code as defined in (9.1).
GetNevOut(blk)	int	Returns the number of activation output ports.
GetNevOutPtrs(blk)	double *	Returns a pointer to the activation output register.

Table 9.3. Activation C macros

Macro	Type	Description
GetNipar(blk)	int	Returns the number of integer parameters.
GetIparPtrs(blk)	int *	Returns a pointer to the integer parameters register
GetNrpar(blk)	int	Returns the number of real parameters.
GetRparPtrs(blk)	double *	Returns a pointer to the real parameters register.
GetNopar(blk)	int	Returns the number of object parameters.
GetOparType(blk,x)	int	Returns the type of object parameters number x.
GetOparSize(blk,x,y)	int	Returns the size of object parameters number x. (y=1 for the first dimension, y=2 for the second dimension)
GetOparPtrs(blk,x)	void *	Returns a pointer to the object parameters number x.
GetRealOparPtrs(blk,x)	double *	Returns a pointer to the real object parameters number x.
GetImagOparPtrs(blk,x)	double *	Returns a pointer to the imaginary part of the object parameters number x.
Getint8OparPtrs(blk,x)	char *	Returns a pointer to the int8 typed object parameters number x.
Getint16OparPtrs(blk,x)	short *	Returns a pointer to the int16 typed object parameters number x.
Getint32OparPtrs(blk,x)	long *	Returns a pointer to the int32 typed object parameters number x.
Getuint8OparPtrs(blk,x)	unsigned char *	Returns a pointer to the uint8 typed object parameters number x.
Getuint16OparPtrs(blk,x)	unsigned short *	Returns a pointer to the uint16 typed object parameters number x.
Getuint32OparPtrs(blk,x)	unsigned long *	Returns a pointer to the uint32 typed object parameters number x.
GetSizeOfOpar(blk,x)	int	Returns the size of the object parameters number x.

Table 9.4. Parameter C macros

Macro	Type	Description
GetNstate(blk)	int	Returns the size of the continuous-time state vector.
GetState(blk)	double *	Returns a pointer to the continuous-time state.
GetDerState(blk)	double *	Returns a pointer to the derivative of the continuous-time state.
GetResState(blk)	double *	Returns a pointer to the residual of the continuous-time state (used only for implicit blocks).
GetXpropPtrs(blk)	int *	Returns a pointer to the continuous-time state properties vector.
GetNdstate(blk)	int	Returns the size of the discrete-time state vector.
GetDstate(blk)	double *	Returns a pointer to the discrete-time state.
GetNoz(blk)	int	Returns the number of object states.
GetOzType(blk,x)	int	Returns the type of object state number x.
GetOzSize(blk,x,y)	int	Returns the size of object state number x (y=1 for the first dimension, y=2 for the second dimension).
GetOzPtrs(blk,x)	void *	Returns a pointer to the object state number x.
GetRealOzPtrs(blk,x)	double *	Returns a pointer to the real object state number x.
GetImagOzPtrs(blk,x)	double *	Returns a pointer to the imaginary part of the object state number x.
Getint8OzPtrs(blk,x)	char *	Returns a pointer to the int8 typed object state number x.
Getint16OzPtrs(blk,x)	short *	Returns a pointer to the int16 typed object state number x.
Getint32OzPtrs(blk,x)	long *	Returns a pointer to the int32 typed object state number x.
Getuint8OzPtrs(blk,x)	unsigned char *	Returns a pointer to the uint8 typed object state number x.
Getuint16OzPtrs(blk,x)	unsigned short *	Returns a pointer to the uint16 typed object state number x.
Getuint32OzPtrs(blk,x)	unsigned long *	Returns a pointer to the uint32 typed object state number x.
GetSizeOfOz(blk,x)	int	Returns the size of the object state number x.
GetWorkPtrs(blk)	void *	Returns a pointer to the Work space.

Table 9.5. States and workspace C macros

Macro	Type	Description
GetNg(blk)	int	Returns the number of zero-crossing surfaces.
GetGPtrs(blk)	double *	Returns a pointer to the zero-crossing vector.
GetJrootPtrs(blk)	int *	Returns a pointer to jroot, the vector that indicates which surfaces have been crossed and in which direction.
GetNmode(blk)	int	Returns the number of modes.
GetModePtrs(blk)	int *	Returns a pointer to the mode register.

Table 9.6. Zero-crossing and mode C macros

In addition to the above macros, the `scicos_block4.h` header file provides utility functions to interact with the simulator in the C computational functions.

- `void do_cold_restart();`
  This function forces the solver to do a cold restart. It should be used in situations where the block creates a non smooth signal. Note that in most situations, non smooth situations are detected by zero-crossings and this function is not needed. This block is used in very exceptional situations.
- `int get_phase_simulation();`
  This function returns an integer which indicates whether the simulator is realizing time domain integration. It can return:
  - **1:** The simulator is on a discrete activation time.
  - **2:** The simulator is realizing a continuous time domain integration.
- `double get_scicos_time();`
  This function returns the current time of the simulation.
- `int get_block_number();`
  This function returns an integer: the block index in the compiled structure. Each block in the simulated diagram has a single index, and blocks are numbered from 1 to `nblk` (the total number of blocks in the compiled structure).
- `void set_block_error(int);`
  Function to set a specific error number during the simulation for the current block. If used, after the execution of the computational function of the block, the simulator ends and returns an error message associated with the number given as integer argument. The following calls are allowed:
  - `set_block_error(-1);`: the block has been called with input out of its domain,
  - `set_block_error(-2);`: singularity in a block,
  - `set_block_error(-3);`: block produces an internal error,
  - `set_block_error(-16);`: cannot allocate memory in block.
- `void coserror(char *fmt,...);`
  Function to return a specific error message in the Scicos editor. If used, after the execution of the computational function of the block, the simulator will end and will return the error message specified in its argument (of type char*).
- `void end_scicos_sim();`
  A very specific function to set the current time of the simulator to the final integration time thus ending the simulation.
  Only expert user should use this function.
- `void * scicos_malloc(size_t);`
  This function must be used to do allocation of Scicos pointers inside a C computational function and in particular when `flag=4` for the work pointer obtained by `GetWorkPtrs`.
- `void scicos_free(void *p);`
  This function must be used to free Scicos pointers inside a C computational function and in particular when `flag=5` for the work pointer.

Examples

The simplest computational functions are those that simply compute an output as a smooth function of an input. One such function is the computational function of the SUMMATION block. The computational function of this block, `summation`, is an old code, which does not take advantage of the C macros. Using the new structure and the proper macros, the computational function of this block can be written as follows:

```
#include <scicos_block4.h>
#include <math.h>

void mysummation(scicos_block *block,int flag)
{
  int j,k,*ipar;
  double *outptr1,*inptr1,*inptrk;
  if(flag==1){
    outptr1=GetOutPortPtrs(block,1);
    if (GetNin(block)==1){
      inptr1=GetInPortPtrs(block,1);
      outptr1[0]=0.0;
      for (j=0;j<GetInPortRows(block,1);j++) {
        outptr1[0]=outptr1[0]+inptr1[j];
      }
    }
    else {
      ipar=GetIparPtrs(block);
      for (j=0;j<GetNin(block);j++) {
        outptr1[j]=0.0;
        for (k=0;k<GetNin(block);k++) {
          inptrk=GetInPortPtrs(block,k+1);
          if(ipar[k]>0){
            outptr1[j]=outptr1[j]+inptrk[j];
          }else{
            outptr1[j]=outptr1[j]-inptrk[j];
          }
        }
      }
    }
  }
}
```

The SUMMATION block can have one or more vector inputs. If it has one input, then the output is a scalar corresponding to the sum of the entries of the unique input vector. If not, the output is a vector of the same size as that of all the input vectors. In this case, the output is simply obtained by the adding and subtracting (in the vector sense, i.e., elementwise) of the input vectors. The choice of addition and subtraction is made based on the values of the integer parameter vector.

The test on the number of inputs determines which algorithm should be used. This number is available from GetNin(blk). The dimensions of the input vector i (counting from one) are given by GetInPortSize(blk,i,1) and GetInPortSize(blk,i,2). The address of the kth input (respectively output) can be found in GetInPortPtrs(blk,k) (respectively GetOutPortPtrs(blk,k)).

Another example of a simple block is the Gain block. This block has a vector input and a vector output. The output is the input multiplied either by a scalar or a matrix. A computational function for this block can be written as follows:

```
#include <scicos/scicos_block4.h>
#include <math.h>
#include <machine.h>
```

```
extern void C2F(dmmul)();

void mygainblk(scicos_block *block,int flag)
{
    int i,un=1,outsz,insz;
    double *inptr1,*outptr1,*rpar;
    inptr1=GetInPortPtrs(block,1);
    outptr1=GetOutPortPtrs(block,1);
    rpar=GetRparPtrs(block);
    insz=GetInPortRows(block,1);
    if (GetNrpar(block)==1){
        for (i=0;i<insz;++i){
            outptr1[i]=rpar[0]*inptr1[i];
        }
    }else{
        outsz=GetOutPortRows(block,1);
        C2F(dmmul)(rpar,&outsz,inptr1,
                   insz,outptr1,&outsz,
                   &outsz,&insz,&un);
    }
}
```

Note that depending on the value returned by `GetNrpar(blk)`, which represents the number of real parameters, two different algorithms are used. If it equals one, then the output is obtained by multiplying each entry in the input vector by the unique real parameter contained in the address given by `GetRparPtrs(blk)`. If not, a matrix multiplication is performed. Note that ScicosLab's multiplication routine is used. `C2F` is used because the routine `dmmul` is in Fortran.

The following is a new version of the computational function associated with the **ABS** block. :

```
#include <scicos/scicos_block4.h>
#include <math.h>

void my_absolute_value(scicos_block *block,int flag)
{
    int i,j,*mode;
    double *outptr1,*inptr1,*g;
    inptr1=GetInPortPtrs(block,1);
    if (flag==1){
        outptr1=GetOutPortPtrs(block,1);
        if(GetNg(block)>0){
            mode=GetModePtrs(block);
            for(i=0;i<GetInPortRows(block,1);++i){
                if (get_phase_simulation()==1) {
                    if (inptr1[i]<0){
                        j=2;
                    } else{
                        j=1;
                    }
                }else {
                    j=mode[i];
                }
```

```
    if (j==1){
      outptr1[i]=inptr1[i];
    } else{
      outptr1[i]=−inptr1[i];
    }
  }
}else{
  for(i=0;i<GetInPortRows(block,1);++i){
    if (inptr1[i]<0){
      outptr1[i]=−inptr1[i];
    }else{
      outptr1[i]=inptr1[i];
    }
  }
}
}else if (flag==9){
  g=GetGPtrs(block);
  mode=GetModePtrs(block);
  for(i=0;i<GetInPortRows(block,1);++i){
    g[i]=inptr1[i];
    if (get_phase_simulation()==1) {
      if(g[i]<0){
        mode[i]=2;
      }else{
        mode[i]=1;
      }
    }
  }
}
}
}
```

This example is not very simple because this block is nonsmooth and thus uses modes. Note, however, that the modes are used only if the block generates continuous-time signals affecting the continuous-time state. The compiler determines whether this is the case, so the computational function must be prepared to function both with and without modes. It is the value returned GetNg(blk) (or equivalently in this case GetNmode(blk)) that determines whether modes should be used. Note that if this value is zero, then a simple absolute value operation is performed in flag 1 and the block is never called with flag 9.

If, on the other hand, this value is not zero, then modes must be used. An important function to use in this case is get_phase_simulation. It returns 1 or 2 to specify the simulation phase. If the numerical solver is at work advancing the time, then the computational function must produce a smooth signal and thus must use the mode to generate its output. When get_phase_simulation returns 1, then the output must be computed normally by computing the absolute value of the input. In the flag 9 case, the zero-crossing surface is computed all the time, but the mode is set only in simulation phase 1.

Note that using modes in this case means that during the simulation, the output of the ABS block can become negative because at zero the solver has to step back and forth for pinpointing the zero-crossing. In some cases, this could be a problem, for example, if the ABS block is followed by a SQRT block that computes the square root. It is for this reason that in the ABS block we have the option of using, or not using, zero-crossings (modes). This is true for most blocks using modes.

evtvardly is the computational function associated with the variable **Event Delay** block in the **Events** palette. The value of the input determines the delay between the activation time of the block and the generation of the output activation. This block is an example of how output activations can be programmed when a block has output activation ports. The following is a new version of the code:

```
#include <scicos/scicos_block4.h>

void my_evt_vardly(scicos_block *block,int flag)
{
  double *inptr1,*evout;
  if (flag==3){
    inptr1=GetInPortPtrs(block,1);
    evout=GetNevOutPtrs(block);
    evout[0]=inptr1[0];
  }
}
```

Let us now consider an example in which the zero-crossing is used to generate a jump in the state of the system. This example corresponds to a bouncing ball. This model includes a gravitational force and a nonlinear air resistance. The differential equations modeling this system are

$$\dot{h} = v, \tag{9.8a}$$
$$\dot{v} = -9.8 - \alpha v^3, \tag{9.8b}$$

where h is the height of the ball, v is its speed, and α is the coefficient of friction. When the ball hits the floor (at $h = 0$), it bounces back without any loss of energy. Thus, at $h = 0$, v becomes $-v$. The computational function of a block realizing this dynamical system is the following:

```
#include <scicos/scicos_block4.h>

void Bounceball(scicos_block *block,int flag)
{
  double *y = GetOutPortPtrs(block,1);
  double *x = GetState(block);
  double *h = &x[0];
  double *v = &x[1];
  double *xd = GetDerState(block);
  double *hd = &xd[0];
  double *vd = &xd[1];
  double *alpha = GetRparPtrs(block);
  if (flag==1){
    *y = *h;
  }else if (flag==0){
    *hd = *v;
    *vd = -9.8-*alpha*(*v)*(*v)*(*v);
  }else if ((flag==2) & (GetNevIn(block)==-1)){
    int *zcd=GetJrootPtrs(block);
    if (*zcd<0){
      *v = -(*v);
    }
```

```
}else if (flag==9){
  double *g=GetGPtrs(block);
  *g = *h;
}
}
```

Note that the test on **nevprt** does not accomplish anything because this function is called only with **flag** equal to 2 when **nevprt** is equal to −1 (**nevprt** indicates the way in which the block is executed and since this block has no activation port and does not inherit any activations because it is always active, it is only internally activated). Note also the test on **jroot**. This test means that we switch the sign of speed only when the ball crosses the zero level going downward. In a perfect world, the ball should never cross zero upward, but due to numerical errors such a situation may come up. In such a case, if a test on the sign of **jroot** is not performed, the ball could bounce off the zero line and go downward thereafter.

To use this function, we should first link it with Scilab. This can be done with the following Scilab commands (assuming that the corresponding file is in the current directory):

```
ilib_for_link('Bounceball','bounceball.o',[],'c')
exec loader.sce
```

The first command compiles the program and creates a shared library. The second command links the shared library with Scilab.

This computational function can be easily tested with the **GENERIC** interfacing function provided in the **Others** palette. The name of the simulation function is set to **Bounceball**, the function type to 4, the initial continuous-time state to **[10;5]**, the real parameter to 0.001, the number of zero-crossings to 1, and the block is declared time-dependent (always active). The block is placed in the diagram in Figure 9.2. The simulation result is shown in Figure 9.3.

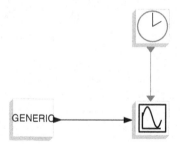

Figure 9.2. Scicos diagram for the bouncing ball example.

Data types

We have seen from the list of macros that the inputs, outputs, some states and some parameters can support various data types. These types are specified in the interfacing function of the block; see Appendix A for details. Table 9.7 gives the data types available in Scicos and their corresponding designation both at the Scilab level, and the C level.

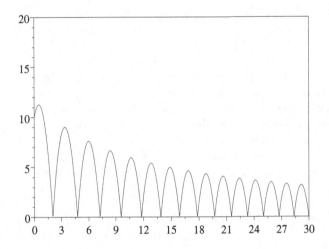

Figure 9.3. Simulation result of the bouncing ball example.

Scilab type	Scicos code	C type	C code
real	1	double	SCSREAL_N
complex	2	double	SCSCOMPLEX_N
int32	3	long	SCSINT32_N
int16	4	short	SCSINT16_N
int8	5	char	SCSINT8_N
uint32	6	unsigned long	SCSUINT32_N
uint16	7	unsigned short	SCSUINT16_N
uint8	8	unsigned char	SCSUINT8_N
boolean	9	long	SCSBOOL_N
other		double	SCSUNKNOW_N

Table 9.7. Scicos and C data types correspondence.

The C codes `SCSREAL_N`, `SCSCOMPLEX_N`, etc., are integers defined in `scicos_block4.h`. For example `SCSREAL_N` is 10. It is however recommended not to use the numbers directly because they may change in the future.

Scilab Computational Functions

A computational function of type 5 is a Scilab function. Its calling sequence is

```
block=func_name(block,flag)
```

where `block` is a Scilab structure (`tlist` of type `scicos_block`) similar to the C structure used in the C computational function of type 4. The fields of `block` are

```
nevprt funpt type scsptr nz z noz ozsz
oztyp oz nx x xd res nin insz inptr nout outsz
outptr nevout evout nrpar rpar nipar ipar nopar oparsz
opartyp opar ng g ztyp jroot label work nmode mode
```

xprop

The fields oz and opar are list of matrices, which can be of different dimensions and types (int8,uint8,etc.).

Note that block.z and block.rpar, in case of computational functions of type 5, can be any Scilab object and not just a vector of reals. The user, however, should make sure that the size of block.z does not change during the simulation. For example, it can be a Scilab list of constant-size vectors, an integer matrix, or a character string, but not a Scilab polynomial because the memory used by a Scilab polynomial depends on the number of nonzero coefficients and is very likely to change if the polynomial changes. block.nz gives the size of the memory used by the Scilab object. So even for a simple vector, block.nz does not give the size of the vector, but rather the size of the vector plus 2. The overhead is due to the coding of the type and dimensions that Scilab places on top of each object. The same applies to block.rpar. In the case of computational functions of type 5, the size information is not really necessary. If block.z is a vector, its size can always be obtained using Scilab's size function:

nz=size(block.z)

The variable flag is a scalar and has exactly the same role as flag in the C case. The following Scilab functions can be used to obtain additional information and perform other operations:

- **curblock()**: returns the current block number in the structure %cpr.
- **scicos_time()**: gives the current time.
- **phase_simulation()**: returns the simulation phase.
- **set_blockerror(i)**: set the error flag if the computational function encounters an error.
- **pointer_xproperty**: used for internally implicit blocks.

Example

The following is a simple type 5 computational function for a Scicos block realizing the sine function:

```
function block=sin5(block,flag)
  if flag==1 then
    for j=1:block.insz(1)
      block.outptr(1)(j)=sin(block.inptr(1)(j));
    end
  end
endfunction
```

Other examples of a type 5 computational functions can be found in Appendix B.1.

Using the Computational Function

Once you have written your computational function, you have to compile it and incrementally link it with Scilab. All this can be done within Scilab. Suppose the name of your C routine is My_prog and it is defined in the file Files.c in the current directory. Then the following Scilab commands can be used to compile and link:

```
⟼    ilib_for_link('My_prog','Files.o',[],'c');
⟼    exec loader.sce
```

These commands work under all operating systems (Linux, Unix, Windows, MacOSX) provided an appropriate C compiler is available on the system.

CBlock4

As we have seen in the previous section, if you are developing a new C computational function, in order to use it in Scicos, you have to compile it and incrementally link it with Scilab before launching Scicos. You also need an interfacing function (or you could use the GENERIC block). The CBlock4 block provides an alternative to this. It allows you to write your computational function on-line. The text is stored in the block structure and the compilation and linking are done automatically. The block also provides a generic interfacing function. The main advantage of using CBlock4 is that the C computational function is included in the block and thus in the diagram structure. This means that the Scicos diagram is self-contained and can be used on different operating systems. The only drawback is that the diagram works only if there is an appropriate C compiler available on the operating system being used. If the computational function is used directly without CBlock4, then dll libraries can be provided along with the diagram and the block will function even if no compiler is available. It also means that the source of the computational function need not be made available. This could be of interest for commercial toolboxes.

9.5.3 Saving New Blocks

New blocks can be used in a Scicos diagram using the AddNewBlock menu by giving the name of the corresponding interfacing function. However, it is a lot more convenient to place new blocks in user palettes. A palette is simply a Scicos diagram from which the user can copy blocks into the Scicos editor. To edit a palette, for example to add a new block, the palette should be loaded as a diagram, edited, and saved.

9.6 Constructing and Loading a New Palette

A palette can be defined if the computational and interfacing functions associated to all of its blocks are available. Suppose we want to construct a palette named foo and we have placed inside the directory /home/basile/foo the C files containing the computational functions foo_blk1 through foo_blk5 and Scilab interfacing functions associated to the blocks to be placed in the palette. Each interfacing function is supposed to be in a file with extension .sci bearing its name.

The C and Scilab files can be compiled and the corresponding libraries generated by the appropriate commands (usually ilib_for_link for C functions and genlib for Scilab functions). These commands can be placed in a file called builder.sce. The following is the content of a builder.sce for compiling the C functions:

```
⟼ comp_fun_lst=['foo_blk1','foo_blk2','foo_blk3','foo_blk4','foo_blk5'];
⟼ c_prog_lst=listfiles('*.c'); //list of C programs in the directory
⟼ prog_lst=strsubst(c_prog_lst,'.c','.o');
```

```
⟼ ilib_for_link(comp_fun,prog_lst,'c'); // compile and generate loader
⟼ genlib('lib_foo',pwd()); // compile macros and generate lib
```

Note that to execute this file, the current directory must be `foo`. The Scilab function `genlib` can be used to construct `lib` associated to Scilab functions as usual. This can be done by the `builder.sce` function as well.

This script needs to be executed only once. It generates the libraries but does not load them. Loading of the interfacing functions can be done with the command `load('lib')` and linking the computational functions done with `exec('loader.sce')`.

At this point, the palette `foo` can be edited and saved in `foo.cos` or `foo.cosf`. Blocks are placed in the diagram using the **Add New Block** functionality.

The palette can also be constructed using the **create_palette** Scilab function. For that, all the files containing the interfacing functions associated with the blocks in the palette should be placed in a directory. Each file must bear the name of the function with the extension `.sci`, as usual. Then the function **create_palette**, called with argument the path to this directory, creates the Scicos diagram of the palette. For example, if the directory `/home/basile/foo` contains the interfacing function files, the command

```
⟼ create_palette('/home/basile/foo');
```

creates a Scicos diagram (palette) `foo.cosf` and places it in `/home/basile/foo`.

To load the palette, we can now construct a `loader.sce` script as follows

```
⟼ load('lib') // loads the interfacing functions
⟼ scicos_pal($+1,1)='foo'; //adding the palette to the list
⟼ scicos_pal($+1,2)='/home/basile/foo/foo.cosf';
```

It is good programming practice to provide with each palette the builder and the loader scripts. These scripts may also be used for creating and loading manual pages for the blocks.

10

Examples and Applications

A typical Scicos model used in a real industrial application includes hundreds of blocks and does not constitute a good example for illustrating Scicos functionalities. That is why in this chapter we start by studying small academic examples for which we can give full details. Later in Section 10.2 we briefly describe a few real-world applications and give references to detailed documentation for interested readers.

10.1 Academic examples

In this section we present a series of small academic examples of modeling and simulation problems implemented in Scicos. Each example is chosen specifically to put forward a different feature of Scicos modeling capabilities.

10.1.1 Predator Prey Model

In this section we consider a simple predator prey model of sardine and shark populations. The dynamics of this model can be expressed as follows:

$$\dot{x} = ax - bxy + F(t)x, \tag{10.1a}$$
$$\dot{y} = cxy - dy, \tag{10.1b}$$

where x denotes the sardine population and y the shark population. The xy terms represent the effect of shark predation of the sardines which is assumed proportional to the frequency of encounters between the two species. These encounters, of course, have a positive effect on the shark population and a negative effect on the sardine population. The a, b, c, and d are positive constants. The $F(t)x$ term represents the consequence of fishing with $F(t) \leq 0$, so that $F(t) = 0$ means no fishing. $F(t)$ may be thought of as a varying harvesting rate.

The Scicos model corresponding to this system of equations is given in Figure 10.1. The fishing effort is modeled by a Super block. With F set to zero, the simulation results are given in Figure 10.2 and Figure 10.3. As we can see, the solution to these equations with $F = 0$ (in fact with F any constant larger than $-a$) is periodic. With the parameters and initial conditions chosen for this simulation, the two populations undergo large variations.

If there is no fishing, then there is a stationary value of this system, which is $x = d/c$, $y = a/b$, obtained by setting the derivatives in (10.1) to zero. Suppose we wish to use fishing regulation to get the sardine population to stay close to the stationary value d/c. One type of regulation, which is practical to implement, is to adjust the fishing effort when

S.L. Campbell et al., *Modeling and Simulation in Scilab/Scicos with ScicosLab 4.4*, 219
DOI 10.1007/978-1-4419-5527-2_10, © Springer Science+Business Media, LLC 2010

populations reach certain threshold levels. So we assume the regulation is done by setting $F(t)$ to 0 (no fishing) when the sardine population is low and $F(t) = f < 0$ when it is high. This is the control law:

$$F(t) = \begin{cases} 0 \text{ if } x \text{ falls below } x_{\min}, \\ f \text{ if } x \text{ goes above } x_{\max}. \end{cases}$$

We set $x_{\min}$ and $x_{\max}$ to 5 percent below and 5 percent above d/c. This control law is implemented in the `fishing effort` Super block as illustrated in Figure 10.4. We could have used the `HYSTERESIS` block to implement this block, but we wanted to illustrate the way `Threshold` blocks function.

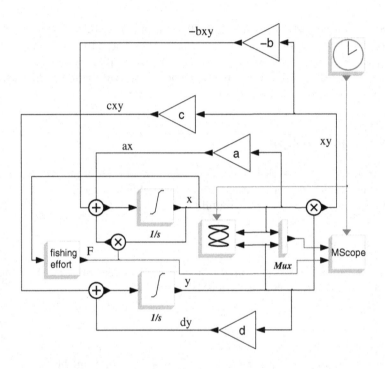

Figure 10.1. The predator prey model with fishing.

The simulation is done for the parameters defined in the context as follows:

```
a=2;b=1;c=.3;d=-1;
perc=5/100;
xmin=-(1-perc)*d/c;xmax=-(1+perc)*d/c;
f=-.5;
```

The simulation results are given in Figure 10.5 and Figure 10.6. As we can see, after a transient period, as desired the sardine population stays much closer to its stationary value than in Figure 10.2.

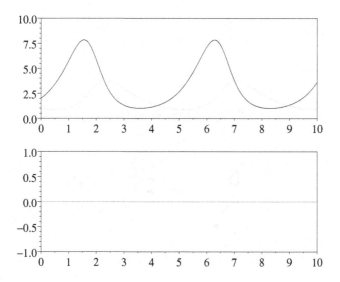

Figure 10.2. The simulation shows wide swings in the two populations for the given initial conditions when the fishing effort (lower graph) is set to zero.

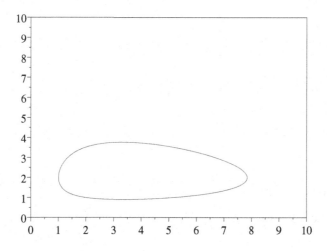

Figure 10.3. The output of the 2D scope displays here the shark population as a function of the sardine population showing clearly the periodicity of the solution.

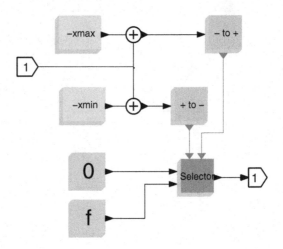

Figure 10.4. Scicos diagram inside the `fishing effort` Super block.

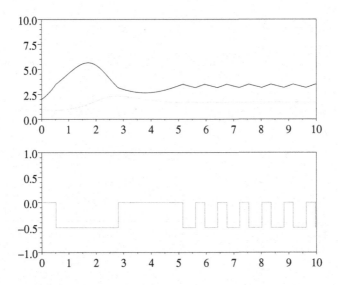

Figure 10.5. The simulation shows the stabilization of the two populations. The lower graph is the fishing effort F.

10.1.2 Control Application

One of the most common uses of Scicos is in control applications. Here we give a simple example in which a discrete-time controller is designed and used to control a continuous-time system.

We start with the observer system constructed in Figure 7.13 of Chapter 7 and construct an observer-based controller. This is done by adding a feedback control to obtain a generic observer-based controller as illustrated in Figure 10.7. The feedback gain Kc is computed as follows:

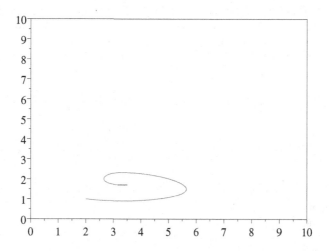

Figure 10.6. The output of the 2D scope shows the solution moving close to the desired stationary point.

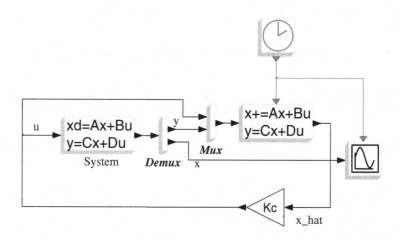

Figure 10.7. Generic Scicos diagram realizing an observer-based controller.

```
Kc=-ppol(A,B,-ones(x0))
```

This statement is, of course, placed in the context of the diagram. In this case, in contrast to what we did in Chapter 7, we can let the matrix A be unstable.

10.1.3 Signal Processing Application

In signal processing applications, very often down-sampling and segmentation are used. These operations can easily be implemented in Scicos. We illustrate this through an example of a simple monitoring scheme, where we track the frequency components of an input signal. We do this by implementing a sliding window over which we apply the FFT (fast Fourier transform) algorithm. The result of the FFT is animated on a graphics window. Moreover, we compute and display on a scope the two dominant frequencies in the signal. Such monitoring is often used in failure-detection applications.

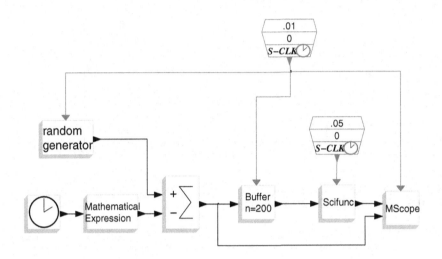

Figure 10.8. A signal processing application.

In the diagram of Figure 10.8, we have used a `Scifunc` block to implement the computation of the FFT and the display of the result. The output of the `Scifunc` block is computed using the following instructions:

```
nz=size(u1,1)/2
f=fft(u1,-1)
f=abs(f(1:nz))
xset('window',3)
xbasc()
plot2d(0:nz-1,f,1,"011"," ",[0,0,nz,240])
[j,I]=gsort(f)
I=I(1:rpar(2))
y1=rpar(1)*(I-1)/(2*nz)
```

Note that we only use half the spectrum. Since the input signal is real, the other half is redundant. Note also that the display of the spectrum is done here using the Scilab graphics function `plot2d`.

The `Mathematical Expression` block is used to generate a test signal as follows:

```
sin(2*%pi*20*u1*(1+cos(u1/20)))+cos(2*%pi*14*u1)*2
```

Here u1 is the block's input, which is just the time. The output is then the sum of a fixed amplitude and frequency cosine and a time-varying frequency sine. Random noise is added to create a more realistic situation.

The `Buffer` is realized using a `CBlock4` block, which allows an on-line definition of its computational function in C. The buffer simply outputs the last n values of its input on its output vector of size n. This is realized as follows:

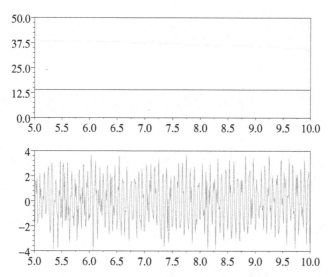

Figure 10.9. Simulation shows the evolution of the two dominant frequencies (top graph) and the original signal (bottom graph).

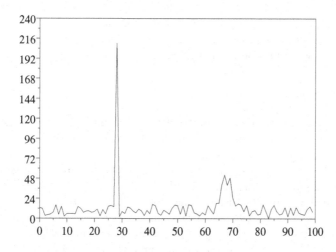

Figure 10.10. The animation of the frequency profile.

```
#include <math.h>
#include <stdlib.h>
#include <scicos/scicos_block.h>
void toto(scicos_block *block,int flag)
{
  if (flag == 4) { /* initialization */
    toto_bloc_init(block,flag);

  } else if(flag == 1) { /* output computation*/
```

```
    set_block_error(toto_bloc_outputs(block,flag));
  } else if (flag == 5) { /* ending */
    set_block_error(toto_bloc_ending(block,flag));
  }
}

int toto_bloc_init(scicos_block *block,int flag)
{
  return 0;
}
int toto_bloc_outputs(scicos_block *block,int flag)
{
  int n=block->outsz[0];
  int i;
  for (i=0;i<n-1;i++){block->outptr[0][n-i-1]=block->outptr[0][n-i-2];}
  block->outptr[0][0]=block->inptr[0][0];
  return 0;
}
int toto_bloc_ending(scicos_block *block,int flag)
{
  return 0;
}
```

The FFT operation can, of course, be implemented in a C block to speed up the simulation. However, in this example, the FFT is not computed at every step, so the overhead of calling the Scilab interpreter is not significant.

The simulation results are illustrated in Figure 10.9 and Figure 10.10. The latter contains a snapshot of the animation at time $t = 20$.

10.1.4 Queuing Systems

Queuing systems can conveniently be modeled in Scicos using the event mechanism. Let us consider a simple queuing system with a Poisson arrival process and an exponential service time. This would model a queue at the checkout counter of a store. Customers arrive following a Poisson process law, and the time each customer spends at the counter is assumed to have an exponential law.

To generate the arrival process, we use the model in Figure 10.11. The random number generator outputs a uniformly distributed random number between 0 and 1. By taking the logarithm and multiplying by $-1/\lambda$, we generate a random number with an exponential law with parameter λ. By feeding the output of the **Event Delay** block to its input activation port, we generate a sequence of events where the time between two events is an independent random variable with exponential law. The result is a Poisson process with parameter λ. The input activation to the Super block starts the process (no event firing is initially programmed on the output of the event delay block).

Generating the departure process (process of serving a customer) is more complicated. As long as the queue is not empty, the time between two events is an independent random variable. But when the queue is empty, no departure event should be generated. The process should be inactive. To implement this process, we have to allow the restart of the process by an external event (the arrival of a customer in an empty queue). This is the reason why the model of the departure events in Figure 10.12 has two input activations. One is for initialization and the other is for restart. Note the use of the **Mathematical**

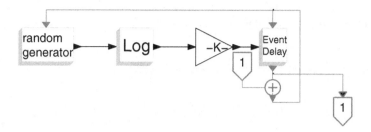

Figure 10.11. Super block generating arrival events based on Poisson's law.

Expression block to allow for modeling more general waiting laws. But in this example we have used `-log(u1)/mu`, which results in an exponential law with parameter μ.

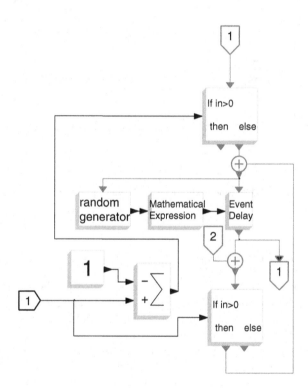

Figure 10.12. Super block generating departure events from the queue.

The complete model is given in Figure 10.13, where the state of the queue is stored in the `1/z` block. Its value goes up or down by one, depending on the event. The simulation result is given in Figure 10.14, where we see the evolution of the queue, and in Figure 10.14, where the events are displayed (dark lines correspond to arrival events). The parameters are defined in the context as follows:

`lam=.3;mu=.35;z0=6`

where z0 is used as the initial state of the 1/z block.

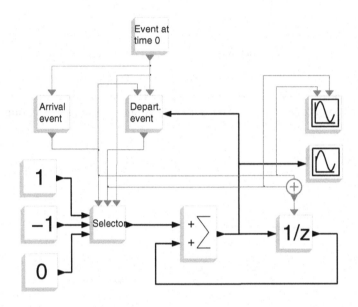

Figure 10.13. Scicos diagram realizing the queuing system.

10.1.5 Neuroscience Application

We consider here a model of the luteinizing hormone-releasing hormone pulse generator. We are not going to discuss the significance of the model. The interested reader is referred to [13]. We focus on the mathematical aspects. This model is chosen because it presents, from the modeling point of view, some nice features.

The model is a continuous-time nonlinear dynamical system with state jumps. The jumps are due to perturbations of variable magnitude occurring at random times. These times are constrained by a minimum interevent period t_a and are modeled as a displaced exponential distribution. The magnitude and timing of the perturbations are modeled as in Figure 10.16. The top random generator block generates uniformly distributed random variables between 0 and $\exp(-t_a\lambda)$, so that the input to the event delay block is a displaced exponential distribution.

The second random generator block generates a normal distribution with mean μ and variance σ^2. The magnitude of the perturbation is then the exponential of this value. Since the perturbation affects only the first state out of the three, two zero signals are concatenated to the perturbation so that the result can directly be added to the state.

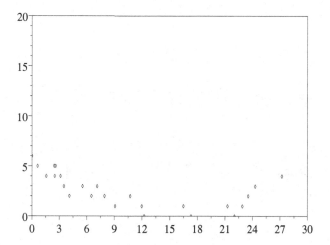

Figure 10.14. Simulation result shows the evolution of the queue.

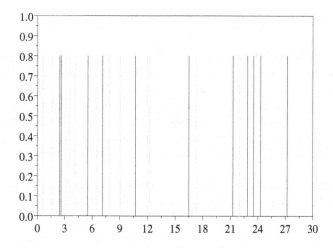

Figure 10.15. Arrival and departure events.

In the absence of perturbation, the dynamics of the system are given by the following equations:

$$\dot{v} = s(-v(v-c)(v-1) - k_1 g + k_2 a), \tag{10.2}$$

$$\dot{g} = b(v)(k_3 a + k_4 v - k_5 g), \tag{10.3}$$

$$\dot{z} = p(v) - k_6 z, \tag{10.4}$$

where

$$b(v) = b_1 - \frac{b_2}{1 + \exp(-b_3(v - b_4))} \tag{10.5}$$

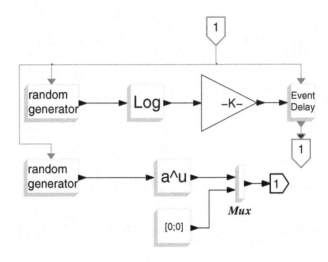

Figure 10.16. Scicos diagram for perturbations.

and

$$p(v) = \frac{p_1}{1 + \exp(-p_2(h(v) - p_3))} \tag{10.6}$$

with $h(v) = v$ if $v > 0$ and $h(v) = 0$ otherwise. The s, a, k_i, p_i and, b_i are known constants.

To allow for a jump in the state, the Jump linear system block is used. In the absence of activation on its input activation port, this block realizes a standard continuous-time linear system with u corresponding to the first input of the block. When the block is activated by an external event, the state jumps to the value specified on the second input. So the second input port must have the same dimension as the state. In this case the state is set to $x = (v, g, z)$ and the parameters are chosen so that the block realizes a simple integrator. This is done by setting $A = 0$, $B = I$, $C = I$, and $D = 0$.

Expressions (10.2) to (10.4) are realized with three Mathematical Expression blocks. See Figure 10.17. Note that the last expression is nonsmooth due to the if statement in the definition of h. This expression is realized in the Mathematical Expression block as follows:

```
(u1>0)*(p1/(1+exp(-p2*(u1-p3)))-k6*u3)+(1-(u1>0))*(p1/(1+exp(p2*p3))-k6*u3)
```

The zero-crossing mode must be set to on.

The simulation result is given in Figure 10.18.

10.1.6 A Fluid Model of TCP-Like Behavior

Our next example is a nonlinear time-delay system representing a fluid model for a TCP/AQM network. Once again we do not address the significance of the model but simply focus on the dynamics and the way it can be implemented in Scicos. We consider this system to show how delay differential systems can be modeled. We use one of the models studied in [34]:

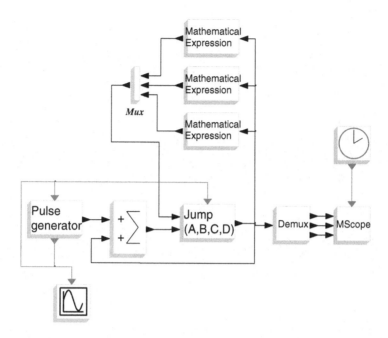

Figure 10.17. Scicos diagram realizing the luteinizing hormone-releasing hormone pulse generator.

$$\dot{W}(t) = \frac{1}{R} - \frac{W(t)W(t-R)}{2R}KQ(t-R), \tag{10.7}$$

$$\dot{Q}(t) = \begin{cases} N\frac{W(t)}{R} - C, & Q > 0, \\ \max(N\frac{W(t)}{R} - C, 0), & Q = 0. \end{cases} \tag{10.8}$$

The Scicos diagram realizing this system is given in Figure 10.19. The continuous-time fixed delay blocks are used to implement the delay. The content of the `Mathematical Expression` block is

`(u2==0)*max(N*u1/R-C,0)+(u2>0)*(N*u1/R-C)`

The simulation result is displayed in the 2D scope, where W is plotted as a function of Q. Note that the result is compatible with that obtained in [43].

10.1.7 Interactive GUI

In some applications, it is desirable to have the possibility of adjusting certain system parameters during the simulation without having to stop and restart the simulation process. We consider here a simple example to illustrate how this can be done. The technique used here can be used to construct sophisticated control panels for controlling simulations.

An on-screen representation of a control that may be manipulated by the user, is called a widget as are parts of such a representation. Scroll bars, buttons, sliders, menu bars, title bars, and text boxes are all examples of widgets. Programmatically, widgets are often expressed as data structures.

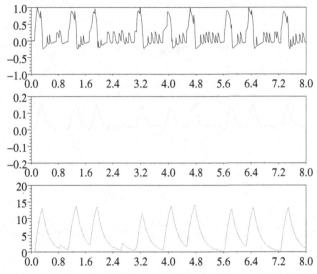

Figure 10.18. Simulation result gives the evolution of the state (v, g, z).

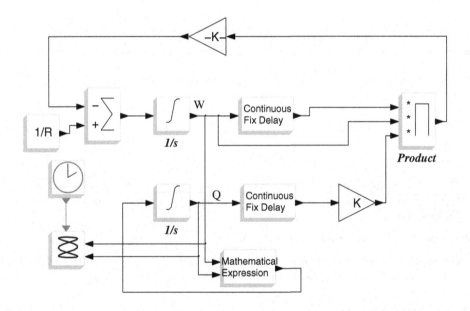

Figure 10.19. The delay differential model.

We consider here a source block where its output can be interactively set using a scale widget by moving up and down a scrollbar. The first thing to consider is the type of widget to use. In order to make sure the application remains machine-independent, a good choice is to use the Tcl/Tk language. Tcl/Tk is included and interfaced in Scilab, so widget

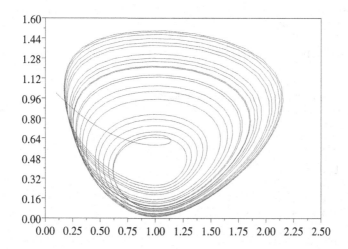

Figure 10.20. Chaotic behavior of the delay differential system.

construction and data exchange can be carried out with Scilab commands. This implies, however, using Scicos blocks of type 5. This may not be a problem, since user interaction is in general of low frequency, and thus the source block can be activated with a slow clock. Otherwise, Tcl/Tk routines must be called directly from the C language, which requires familiarity with the internal implementation of Tcl/Tk, to have optimal performance.

The source block we are considering must construct and display the widget at initialization (flag 4) and then read the position of the scrollbar and return it when activated with flag= 1. All this is done, of course, by the computational function, which here will be of type 5, that is, written in Scilab.

The interfacing function is the following:

```
function [x,y,typ]=TKSCALE(job,arg1,arg2)
  //Source block; output defined by tk widget scale
  x=[];y=[];typ=[];
  select job
    case 'plot' then
    standard_draw(arg1)
    case 'getinputs' then
    [x,y,typ]=standard_inputs(arg1)
    case 'getoutputs' then
    [x,y,typ]=standard_outputs(arg1)
    case 'getorigin' then
    [x,y]=standard_origin(arg1)
    case 'set' then
    x=arg1;
    graphics=arg1.graphics;exprs=graphics.exprs
    model=arg1.model;
    [ok,a,b,f,exprs]=getvalue('Set scale block parameters',..
        ['Min value';'Max value';'Normalization'],..
        list('vec',1,'vec',1,'vec',1),exprs)
    // tk widget returns a scalar, the value is divided by
```

```
    // Normalization factor
    if ok  then
      graphics.exprs=exprs
      model.rpar=[a;b;f]
      x.graphics=graphics;x.model=model
    end
    case 'define'  then
    a=0;b=100;f=1;// default parameter values
    model=scicos_model()
    model.sim=list('tkscaleblk',5)
    model.out=1
    model.evtin=1
    model.rpar=[a;b;f]
    model.blocktype='d'
    model.dep_ut=[%f %f]
    exprs=[sci2exp(a);sci2exp(b);sci2exp(f)]
    gr_i=['xstringb(orig(1),orig(2),''TK Scale'',sz(1),sz(2),''fill'')']
    x=standard_define([3 2],model,exprs,gr_i)
  end
endfunction
```

Note that the block has three parameters: the first two are the minimum value and the maximum value of the scale. The last is a normalization factor. The output of the block is normalized by dividing the scale value by this factor. This allows, in particular, the display of only integer scale values inside the widget.

The computational function is the following:

```
function block=tkscaleblk(block,flag)
  blknb=string(curblock())
  if flag==4  then
    cur=%cpr.corinv(curblock())
    if size(cur,'*')==1  then // open widget only  if the block
                              // is in main Scicos editor window
      o=scs_m.objs(cur).graphics.orig;
      sz=scs_m.objs(cur).graphics.sz
      pos=point2pixel(1000,o)
      pos(1)=pos(1)+width2pixel(1000,sz(1)) // widget position
      geom='wm geometry $w +'+string(pos(1))+'+'+ string(pos(2));
      title=block.label
       if title==[]  then title="TK source", end
      tit='wm title $w Scale'+blknb // write block label
      bounds=block.rpar(1:2)
      bnds='-from '+string(bounds(1))+' -to '+string(bounds(2))
      cmd='-command ""f'+blknb+' $w.frame.scale""'
      lab='-label ""'+title+'""';
      L='-length 100'
      I='-tickinterval '+string((bounds(2)-bounds(1))/4)
      scale=strcat(['scale $w.frame.scale -orient vertical',..
                    lab,bnds,cmd,L,I],' ')
      initial=mean(bounds) // initial value is the mean
      txt=['set w0.vscale'+blknb;
           'set y'+blknb+' 0';
           'catch destroy $w';
```

```
           'toplevel $w';
           tit
           geom
           'frame $w.frame -borderwidth 10';
           'pack $w.frame';
           scale
           'frame $w.frame.right -borderwidth 15';
           'pack $w.frame.scale -side left -anchor ne';
           '$w.frame.scale set '+string(initial);
           'proc f'+blknb+' w height global y'+blknb+';set y'+blknb+' $height'
           ];
      TCL_EvalStr(txt) // call TCL interpretor to create widget
      block.outptr(1)=mean(block.rpar(1:2))/block.rpar(3);
    end
  elseif flag==1  then // evaluate output during simulation
    block.outptr(1)=evstr(TCL_GetVar('y'+blknb))/block.rpar(3);
  end
endfunction
```

The block number (returned by the function `curblock`) is used to customize Tcl variable names to make sure that if two of these blocks are used in the same diagram, they create different widgets. The position of the widget is computed so that it is placed next to the corresponding block in the Scicos diagram (it can be moved subsequently by the user). The complete Tcl text for the construction of the widget is sent to the Tcl interpreter using the `TCL_EvalStr` command. The Tcl procedure defined in the program sets the variable yN (where N is the block number) to the value designated by the position of the scrollbar in the scale. This variable is returned from the Tcl environment to Scilab using the command `TCL_GetVar` in the case of flag= 1. It is then normalized and returned as the output of the block.

The source block `TKSCALE` can be used, for example, to adjust manually the parameters of a PD controller. This is done in the diagram illustrated in Figure 10.21. The clock period activating the square-wave generator is set to half that of the refresh period of the scope so that the output of the generator, displayed as the first curve on the scope, remains always the same, that is, -1 up to the middle of the screen and then 1. The output of the plant is changed as the user moves up and down the scrollbars in the GUI boxes. See the simulation result in Figure 10.22.

Various Tcl/Tk widgets can be used to display simulation results and even to construct a scope using Tk Canvas. It is not difficult to imagine the construction of sophisticated control panels, for example, the instrumentation panel of a car in Tcl/Tk to be used with a simulation model of the car. This can be done even if the model is simulated in batch mode.

10.2 Applications

Scicos has been used in many industrial applications that we know of, and certainly in a lot more that we are not aware of. Our objective here is not to provide an exhaustive list of applications but simply to list a few typical applications of Scicos without getting into the details. When possible, we provide links to where additional information, and sometimes even the Scicos code, can be downloaded.

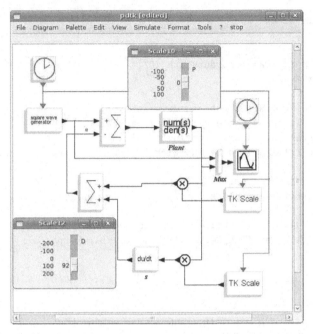

Figure 10.21. Screen capture of the diagram shows the widgets during simulation.

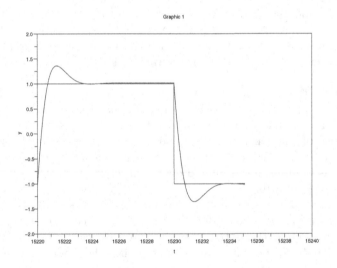

Figure 10.22. Simulation result after adjusting PD parameters by hand using the widgets.

Automative industry

Scicos has been used in the automotive industry for a long time. One of its first industrial applications has been the modeling of a lean burn engine with direct injection for a major European car manufacturer. The hybrid modeling capabilities of Scicos, in particular the ability to control explicitly the events, has proved to be very useful in this application where the model included the coordination between the automatic transmission mechanism and the engine control. This coordination allows the controller to take advantage of the gear shifting decision in order to anticipate the changes required in the torque to assure a smooth ride during the actual gear shifting. The gear shiftings were naturally modeled as Scicos events and the engine was modeled as a hybrid system. The model was used to optimize and validate a control law.

Engine control is one of the most important application domains for Scicos; see for example [44, 48]. But Scicos has also been used in other automotive applications, such as trajectory control [17].

Aeronautics

Another industry that relies heavily on modeling and simulation tools, such as Scicos, is the aeronautics industry. Scicos has been used for modeling thrust reversers on civilian airplanes. Thrust reversers are used during landing to help slow down the airplane.

Scicos has also been used for modeling oxygen masks. A pilot's oxygen mask is a complex mechanical device that includes a feedback mechanism to regulate the pressure. The model has been used for adjusting design parameters to satisfy robustness specifications and to optimize the comfort of the pilot (in particular, to reduce vibrations when inhaling). The use of Scicos for this application actually showed its limitations in modeling mechanical devices and became the principal motivation for introducing implicit models into Scicos.

Energy production

The first large industrial application of Scicos has been for the design, the analysis, and the simulation of the operation of hydroelectric equipment (hydroelectric power plants, reservoir or river dams, etc.). This work has been realized by the Research and Development Division of EDF (Electricity of France). See [25] for details.

Scicos, and in particular its Modelica components, are also being used for modeling the cooling system of nuclear power plants. The physical laws, mainly from thermodynamics, are modeled in Modelica, and the control laws are implemented in classical Scicos.

Scicos is used in many other applications in this domain; see for example [45].

Engineering education

The free open-source nature of Scilab/Scicos has played an important role in its widespread use in education. Scilab/Scicos is available today in most French universities and engineering schools, as well as many universities around the world. Scilab/Scicos provides a complete environment for use in systems, control and signal processing courses. Such an environment is not available in any other free open-source software.

Scicos is particularly used in control laboratories for teaching controller design. There have been a number of laboratory experiments developed that allow students to implement and evaluate feedback strategies with Scicos, either using a hardware in the loop approach or by code generation. See for example [42] where a rotary inverted pendulum is considered. The classical ball and beam experiment is discussed in [39, 40]. More information can be found on http://www.scicos.org/scicoshil.html.

Telecommunication

Scicos is particulary adapted for modeling signal processing and communication algorithms. In addition, the Modnum toolbox provides many advanced functionalities: http://www.scicos.org/ScicosModNum/modnum_web/web/eng/eng.htm. Scicos is used in different areas of telecommunication. For example, it is extensively used by a French company for the design of wireless handset applications.

Realtime applications

Once a controller or a filter has been designed and validated in Scicos, often realtime code generation is needed for the actual implementation. A number of companies provide products and services for realtime code generation based on Scicos such as Evidence in Italy (http://www.evidence.eu.com), CosateQ in Germany, Industrial Automation Freedom in Switzerland (http://www.iafreed.com), etc.

Other applications

There are various other application domains in which Scicos has been used. In medical research, the toolbox LARY-CR has been used in a number of studies [41]; see for example [38]. There have been applications in power electronics, with and without Modelica components. RTSS, a toolbox for Robotics, has been presented at the HeDiSC Workshop on Open Source Software for Control Systems, which took place in Lugano, Switzerland, in 2009. We have encountered applications for studying queuing chains using Monte Carlo techniques, and other statistics based studies, in particular in Finance.

11

Batch Processing in Scilab

In ScicosLab, most Scicos functionalities are developed in the Scilab language. This is very useful because it allows Scilab and Scicos to interact in many different ways, thereby mutually enriching their functionalities. We have already seen that Scicos relies heavily on the Scilab language. The Scicos editor is, for example, fully written in Scilab, making it easy for the user to add new functionalities or customize existing ones. We have also seen that the interfacing functions of all Scicos blocks are Scilab functions using only standard Scilab commands. For example, to define a block's icon we use standard Scilab graphics commands. But perhaps the most important aspect of having access to Scilab in Scicos is the ability to use Scilab commands from specialized toolboxes, such as Signal Processing, Control and Optimization, in the construction of Scicos blocks.

In this chapter we look at another important area where the Scilab Scicos interaction is used. We describe in particular the ways Scicos simulations can be performed and used in the Scilab environment. We have already seen that Scilab can be used for postprocessing Scicos simulation results, but here we look at ways of piloting Scicos functionalities through Scilab commands.

In doing modeling and simulation work, very often one needs to perform multiple simulations either for adjusting or optimizing model parameters, or for testing the model on multiple data sets, or for performing simulation-based statistical analyses. Monte Carlo techniques are one example. In such cases, Scicos simulations should be carried out in batch mode. In the Scilab/Scicos environment, this can be done within the Scilab language, which not only can handle repetitive calls to the Scicos simulator by changing the value of a parameter or the name of a file each time, but can also perform complicated analyses and make decisions on how to proceed with additional simulation runs.

11.1 Piloting Scicos via Scilab Commands

Having the Scicos editor written entirely in Scilab simplifies a great deal the piloting of Scicos commands from Scilab. Note, however, that the Scicos library is not loaded by default into the Scilab environment. So if you want to use any Scicos routines within Scilab, you must first load them as follows:

```
⊢⟶   load SCI/macros/scicos/lib
```

S.L. Campbell et al., *Modeling and Simulation in Scilab/Scicos with ScicosLab 4.4*,
DOI 10.1007/978-1-4419-5527-2_11, © Springer Science+Business Media, LLC 2010

if you are writing a Scilab function, you can place this instruction on top of it. Scicos blocks' interfacing functions associated with the blocks inside Scicos palettes are not loaded automatically either. If they are needed, they can be loaded as follows:

```
⟼    exec (loadpallibs,-1)
```

You can now use Scicos commands associated with each editor button. The name of the function is obtained from the text inside the button from which empty spaces, slashes, dots and dashes have been removed and an underscore has been added to the end. For example, the Scilab function associated with the button Save As is named SaveAs_. It is defined in the directory scilab/macros/scicos, like all other editor functions, and inside the file SaveAs_.sci.

Note, however, that using editor commands requires a good understanding of the data structures and global variables used in the Scicos editor. Some of these functions require that these variables be defined properly. In order to see how the editor functions should be called, we examine the scicos function. This function, which can be found in scilab/macros/auto/scicos.sci, is the main Scicos editor program and contains the main loop from which editor functions are called when the Scicos editor is active. Some data structures used in this function are explained in Appendix A.

In many applications, however, it turns out that there is no need to access directly every editor command. The reason is that often when doing batch processing, we are only interested in being able to change system parameters and running the simulation with these new parameters. The Scilab function scicos_simulate provides a generic way of performing batch simulation. This function is described in Section 11.1.2.

But first we describe scicosim, the basic Scilab function interfacing the Scicos simulator. To modify model parameters when using directly scicosim requires a manual adjustment of the result of the compilation. This is not straightforward and requires an understanding of the way the compilation result is stored. Using this direct method, not only does not allow the usage of some Scicos facilities such as the From Workspace and tt To Workspace blocks but also has the disadvantage that Scicos cannot check the result of the modification to assure its consistency. Some modifications may lead to inconsistent compiled Scicos structures that can crash the simulator. Others, may lead to erroneous simulations. That is why in most cases it is recommended to use the Scilab function scicos_simulate. When the modification consists simply in changing the parameter values of blocks, the direct use of scicosim can be envisaged because using scicosim has practically no overhead and thus can be more efficient than using the function scicos_simulate. Note however that even in such a case, some parameter modifications may alter blocks' fundamental properties thus requiring a recompilation. Consider for example the linear system block described in terms of the system matrices A, B, C and D. The matrix D being zero or not determines the direct feedthrough property of the block. This property is a key block property used by the compiler. So if the parameter modification consists in changing a zero D into a non-zero D, the model has to be recompiled, which means that scicosim cannot be used directly. The function scicos_simulate automatically determines the need for recompilation and performs it in this case.

11.1.1 Function scicosim

The Scilab function scicosim is an interface to the Scicos simulator. To use this function, the Scicos diagram must be compiled because we need the compilation result %cpr. The user, however, need not be familiar with the compilation stage (information on that can be found in Appendix A).

When used for simulation, `scicosim` is called in three different ways. First it is called to initialize the simulation. This is done usually by the following Scilab command:

```
[state,t]=scicosim(state,0,tf,sim,'start',tol)
```

The string `start` indicates that only the initialization must be performed. To run the simulation and finish the simulation, the following corresponding calls are performed:

```
[state,t]=scicosim(state,t,tf,sim,'run',tol)
```

and

```
[state,t]=scicosim(state,t,tf,sim,'finish',tol)
```

The argument `state` initially contains the initial state of the system to be simulated (see Section A.2.3 for details). It contains, in particular, initial values of all the continuous-time and discrete-time states of all the blocks, the preprogrammed events, etc. `state` is both an input and an output because it changes as the simulation goes on.

The argument `sim` contains the scheduling tables needed by the simulator. These tables are generated by the compiler and do not evolve during the simulation (see Section A.2.3 for details).

The argument `tf` is the time up to which the simulation must run (not used with options `start` and `finish`). When `scicosim` is called with the `run` option, the simulation is performed up to time `tf` unless a pause or an error is encountered earlier. The output argument `t` gives the actual ending time of the simulation. In most cases we have `t` equal to `tf`. The input argument `t` indicates the current time and is used only in the `run` case. Its value is, in general, 0 because Scicos simulations by default start at time 0. But if a simulation is to be continued by making other calls to `scicosim` with option `run` without starting over, then the final time of the previous simulation (the output argument `t` of previous call to `scicosim`) must be used.

The argument `tol` is a vector containing various simulation parameters. It includes

1. `atol`: absolute tolerance (see Section 3.2.1).
2. `rtol`: relative tolerance (see Section 3.2.1).
3. `ttol`: minimum spacing in time between two events for the continuous-time solver to be called. If two events are spaced closer than `deltat` in time, the solver is not called and the continuous-time variables are supposed to remain constant over that period.
4. `deltat`: maximum time interval length used for each call to the continuous-time solver. If the next event is further in the future than `deltat`, the continuous-time solver is called more than once.
5. `scale`: sets the correspondence between simulation time and real time. If set to 1, one unit of Scicos simulation time corresponds to one second. The solver is slowed down to achieve the desired correspondence. The default value 0 indicates that no slowing down is to be performed and simulation runs at maximum speed.
6. `hmax`: maximum time step taken by the solver (see Section 3.2.1). The default value 0 indicates no maximum being set.

To use `scicosim`, the variables `state` and `sim` must be available. The initial value of `state` and the value of `sim` are computed by the Scicos compiler from the Scicos diagram. It is possible to call the compiler from Scilab to construct these variables, but the simpler solution is to use the Scicos editor to construct them and save them to a file for future use in Scilab. This can be done as follows:

1. Use the Scicos editor to construct a block diagram and compile it.
2. Save the compiled diagram in binary form (as a `.cos file`) using the **Save** or **Save As** menu.
3. In Scilab, load the saved file (say `myfile.cos`) using the `load` function:

   ```
   load myfile.cos
   ```

   You obtain Scilab variables `scicos_ver`, `scs_m`, and `%cpr`. `scicos_ver` contains the version number of Scicos that has created the diagram, `scs_m` contains the Scicos diagram, and `%cpr` contains the result of the compilation. It is this latter that is of interest to us for calling `scicosim`. In particular, `%cpr` is a Scilab structure containing `state`, `sim`, `cor`, and `corinv`. Note that if the diagram is not compiled before being saved, then `%cpr` is an empty structure.
4. Extract `state` out of `%cpr`:

   ```
   state=%cpr.state;
   ```

5. Define `tol` and `tf`, for example,

   ```
   tf=1;
   tol=[1.d-4,1.d-6,1.d-10,tf+1,0,0];
   ```

   then call `scicosim` for initialization:

   ```
   [state,t]=scicosim(state,0,tf,%cpr.sim,'start',tol);
   ```

6. Then run the simulation from 0 to `tf`:

   ```
   [state,t]=scicosim(state,t,tf,%cpr.sim,'run',tol);
   ```

   Successive such calls may be performed changing the final time `tf` before each call.
7. End the simulation:

   ```
   [state,t]=scicosim(state,t,tf,%cpr.sim,'finish',tol);
   ```

Running Scicos simulations in Scilab in batch mode is useful if the user wants to perform multiple simulations by changing a parameter or an initial condition each time. All the parameters of the blocks in the diagram are in `%cpr.sim`. The initial conditions are in `state`. Note, however, that the change in the parameter should not require recompilation. For example, if one is testing the performance of a controller by changing its parameters, the order of the controller should not be modified, because this implies a change in the size of a state, which in turn requires a partial recompilation since `%cpr` is no longer valid.

To change the parameter of a block, one must know the block number in the compiled structure. This information can be obtained in the Scicos editor using the **Get Info** menu in the "Compiled structure index" field. This number specifies the order in which blocks are placed in `%cpr` fields. Note, however, that this number may change if the diagram is edited and recompiled.

This number can also be obtained in Scilab if a "label" has been associated with the block of interest. Labels are available in `%cpr.sim.labels` as a vector. So, for example, if the label `toto` has been assigned to a block, the block number can be obtained using the command

```
k=find(%cpr.sim.labels=='toto')
```

k is then the block number and can be used to examine and modify block parameters. Suppose the block labeled `toto` is a `GAIN` block. The parameters of the block that are the entries of the gain matrix, like all real parameters of all the blocks, are in `%cpr.sim.rpar`. The entries of the vector `%cpr.sim.rpar` that contain the parameter of block k can be obtained from `%cpr.sim.rpptr`, which is a vector of pointers to `%cpr.sim.rpar`. In particular, let

```
idx=%cpr.sim.rpptr(k):%cpr.sim.rpptr(k+1)-1
```

Then `idx` contains the indices of `%cpr.sim.rpar` corresponding to the `GAIN` block:

```
gain_param=%cpr.sim.rpar(idx)
```

So if the objective is to run simulations for various gains, then before starting each simulation, the parameter must be set as follows:

```
%cpr.sim.rpar(idx)=new_gain_param
```

To change integer parameters, the procedure is identical except that the two vectors to use are `%cpr.sim.ipar` and `%cpr.sim.ipptr`. For initial continuous-time and discrete-time states, the pointer vectors are respectively `%cpr.sim.xptr` and `%cpr.sim.zptr`, and the values are found in `%cpr.state.x` and `%cpr.state.z`.

Example of Batch Simulation

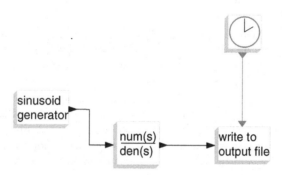

Figure 11.1. Scicos diagram for batch simulation.

Consider the Scicos diagram illustrated in Figure 11.1. The sinusoid generator block has been labeled `toto`, the transfer function in the linear system block is $1/(1+s+s^2)$, and the output file name is chosen to be `o_file`. Suppose we want to simulate this diagram for different values of the frequency parameter in the **sinusoid generator** block and compare the results in Scilab by plotting them in the same plot. This can be done by changing the second parameter of the **sinusoid generator** (the three parameters of this block are respectively the amplitude, the frequency, and the phase) before each simulation. Note that the file name is not changed from one simulation to the other, so at the end of each simulation the file `o_file` must be stored in a different location. The following Scilab script does exactly that if `batch1.cos` has been saved *after* being compiled.

```
load batch1.cos
freq_list=[.1 1 2 4];                          ← list of frequencies to consider
k=find(%cpr.sim.labels=='toto');                        ← block number
idx=%cpr.sim.rpptr(k):%cpr.sim.rpptr(k+1)-1;
tf=100;                                         ← simulation parameters
tol=[1.d-4,1.d-6,1.d-10,tf+1,0,0];
 for i=1:size(freq_list,2)                      ← main loop over the list of frequencies
   %cpr.sim.rpar(idx(2))=freq_list(i);
   state=%cpr.state;
   [state,t]=scicosim(state,0,tf,%cpr.sim,'start',tol);
   [state,t]=scicosim(state,t,tf,%cpr.sim,'run',tol);
   [state,t]=scicosim(state,t,tf,%cpr.sim,'finish',tol);
   // renaming o_file o_file1, o_file2,....
   if MSDOS  then
     host('move o_file o_file'+string(i));
   else
     host('mv o_file o_file'+string(i));
   end
 end
// and plotting on a single graph
xbasc()
 for i=1:size(freq_list,2)
   tx=read('o_file'+string(i),-1,2);            ← reading back the results of the simulation
   plot2d(tx(:,1),tx(:,2),i)                    ← and plotting on a single graph
 end
```

In this example, we can avoid using the **host** command by changing the file name in the **write to output file** block. The file name is coded as an integer in the integer parameter vector of the block. To do so, however, we must know exactly where the name appears in the vector. This information can be obtained by examining the interfacing function of the block. The integer coding of the name can be done using the Scilab function **str2code**. The function retrieving the name from the code is **code2str**. Note, however, that if we are to use this method to write into different files, all file names must have the same length, unless **%cpr.sim.ipptr** is reconstructed. This is not recommended, especially since there is a more convenient way of handling such modifications, as we shall see in the next section.

Finally, note that in this example the diagram contained no graphics displays such as a scope. This is often the case when one is doing batch processing. But sometimes it could be interesting to follow simulation results visually for debugging purposes.

11.1.2 Function scicos_simulate

The **scicos_simulate** function provides a convenient way of running Scicos simulations in Scilab using the diagram's data structure **scs_m**. Unlike **scicosim**, **scicos_simulate** does not require **%cpr**. The function **scicos_simulate** handles automatically all the necessary compilations in a transparent way.

In this case, model changes must be done by changing formal parameters defined normally in the **context**; see Section 11.2.1 for information on formal parameters and the use of **context**. Changing formal parameters not only is more convenient than directly changing block parameters, it also allows for more general changes. For example, it is more natural in the case of a linear system block, specified in terms of a transfer function, to specify the new polynomials representing the transfer function's numerator and denominator than it is to change directly the matrices of the state-space representation. Moreover, by changing the transfer function, the order of the linear system and thus the size of its system matrices can vary.

In this case the `scicos_simulate` function performs automatically the necessary (partial) recompilation.

The calling sequence of this function is as follows:

```
Info=scicos_simulate(scs_m,Info[,%scicos_context][,flag]);
```

The arguments `%scicos_context` and `flag` are optional.

- `scs_m`: scicos diagram, obtained for example by the instruction `load mydiag.cos`. Note that the diagram `mydiag` need not be compiled because we do not make use of `%cpr`.
- `%scicos_context`: a Scilab structure containing values of symbolic variables used in the context and Scicos blocks. This is often used to change a parameter in the diagram context. In that case, in the diagram context, the variable must be defined such that it can be modified. Say a variable `a` is to be defined in the context having value 1, and later in batch mode it is to be changed. In that case, in the context of the diagram one can use the following command to define `a`:

```
if ~exists('a') then a=1,end
```

This way we define the default value of `a` to be 1 but we allow its modification. For example, to run a simulation in batch mode using the value a= 2, it suffices to set `a` to 2 in the `%scicos.context` variable. This can be done as follows:

```
%scicos_context.a=2
```

- `Info`: a list. It must be `list()` at the first call, then output `Info` can be used as input `Info` for the next call. `Info` contains compilation and simulation information and is used to avoid recompilation when not needed. Normally the user should not, and need not, modify `Info`. There are, however, a few exceptional cases in which expert users may be able to edit this variable. `Info` contains the following fields (when it is not empty):
  1. `%tcur`: starting time of simulation. It is always zero except if the simulation had been halted before reaching the end time, for example by an event to the block `STOP`. In this case the simulation is not terminated but only halted at time `%tcur`.
  2. `%cpr`: compilation result as we have previously seen.
  3. `alreadyran`: a boolean indicating whether the simulation is halted.
  4. `needstart`: a boolean indicating whether the simulation is to be restarted from zero. This applies only if the simulation is halted. If false, the halted simulation proceeds without reinitialization.
  5. `needcompile`: An integer indicating the level of compilation. It should always be zero unless there are errors in the diagram leading to the failure of the compilation process. But it can be used by the user to force a compilation. It should be set to 4.
  6. `%state0`: Initial state.
- `flag`: a string. If it equals `nw` (no window), then blocks using graphics windows are not activated. In most batch simulation tasks, graphical outputs are not useful and only slow down the simulation. Note that the list of deactivated blocks must be updated as new blocks are added. The current list is coded in the following vector:

```
['cscope','cmscope','scope','mscope','scopexy','evscpe','affich']
```

We shall see examples of the usage of `scicos_simulate` in Sections 11.2.2 and 11.2.3.

11.2 Data Sharing

When running a Scicos simulation in Scilab, exchanging data between Scilab and Scicos becomes an important issue. In previous examples we have seen a few methods. In particular, we have seen how to communicate information from Scilab to Scicos using the `%scicos_context` variable. We have also seen how Scicos simulation results can be retrieved in Scilab through writing into a file. We shall examine in more detail these methods and propose new ones for exchanging data.

11.2.1 Context Variables

The variable `%scicos_context` is used to store all the variables defined in the context. As we have seen in previous chapters, Scicos blocks' parameters can be made to depend on formal parameters. These formal parameters (which are just Scilab variable names) must be defined in the context of the diagram before being used. All the formal variables defined in a context are placed by the Scicos editor in the Scilab structure `%scicos_context`. So, for example, if the context contains the instruction

```
a=3
```

then the field `a` is added to `%scicos_context` as if the instruction

```
%scicos_context.a=3
```

had been executed. After that, `a` can be used in block dialog boxes to set block parameters. If a variable is not defined in the context, then it cannot be used, unless it is placed initially in `%scicos_context`. The way context works is to first extract all the variables from `%scicos_context` and then execute context commands and finally place all the variables defined by these commands in `%scicos_context`. Variables already in `%scicos_context` can thus be used and updated, if needed, by the context.

In most cases, when one is using the Scicos editor, `%scicos_context` is initially empty, and all formal parameters must be defined by the context commands. But a user can initialize `%scicos_context` before launching Scicos (editor) by placing certain parameters in it. The user can then use these parameters without having to define them in the context. The user just should make sure that the initialization is done every time Scicos is started.

But the real usage of `%scicos_context` occurs when the Scicos simulator is to be executed by a Scilab command. Being able to change formal parameters is an extremely flexible way of changing system parameters. We shall see that in Sections 11.2.2 and 11.2.3 through examples.

11.2.2 Input/Output Files

Another way of exchanging data between Scilab and Scicos is through the usage of external files. We have seen an example of that previously. Files are particularly useful if the data to be exchanged is not just a value but a timed sequence of values. For example, an input signal to a Scicos diagram performing signal processing, is best placed in a file. Similarly,

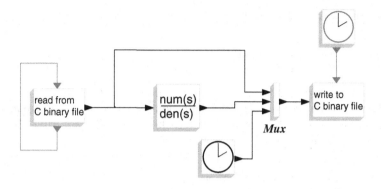

Figure 11.2. Using files for exchanging data between Scicos and Scilab.

simulation results in the form of output signals can be placed in output files. These files can then be read into Scilab for further processing.

A simple example of such exchange is the following. Consider the Scicos diagram in Figure 11.2. In this diagram, the **read** block generates a piecewise constant signal by reading the values and the timing of the signal from the file **foo**. The dialog box of this block is illustrated in Figure 11.3.

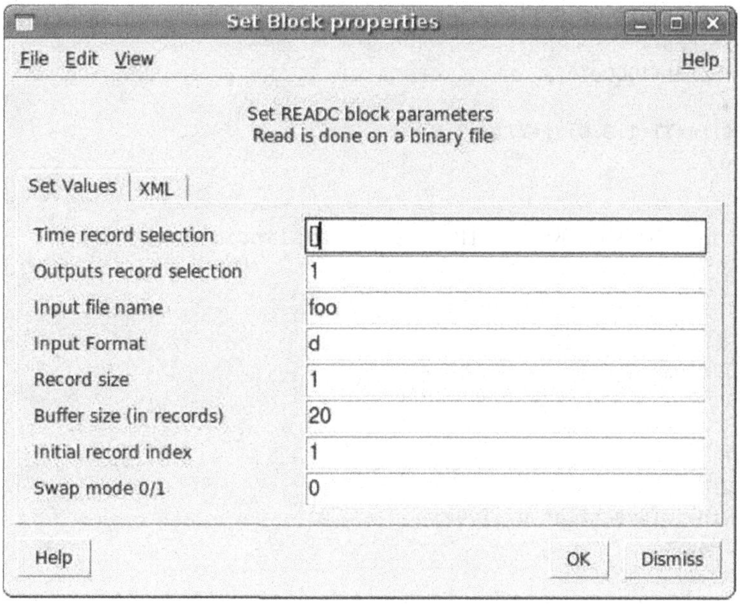

Figure 11.3. Parameters of the **read** block.

The signal generated by the **read** block goes through a continuous-time linear system the output of which, along with the input and the current time, are fed to a **write** block that writes the data to the file **goo** at the frequency specified by the parameters of the event clock. The "period" of this latter is set to dt defined in the context along with the parameters of the linear system as follows:

```
if ~exists("num") then num=1,end
if ~exists("den") then den=1+%s,end
if ~exists("dt") then dt=0.1,end
```

In order to run this diagram in batch mode from Scilab, we define a function which receives the description of the input signal and the definition of the linear system, generates the file foo, sets the parameters of the linear system in the Scicos diagram, runs the simulation, reads the content of the file goo, and outputs it:

```
function [t,u,y]=syslinsim(Sys,tf,dt,ui,T,U)
 // Define symbolic parameters
 %scicos_context.num=Sys.num
 %scicos_context.den=Sys.den
 %scicos_context.dt=dt
 // Add initial input and a fake  end point
 UT=[[0,T,tf+1];[ui,U,0]];
 mopen("foo","wb")
 mput(UT,"d")
 mclose()
 load("bg.cos")
 scs_m.props.tf=tf
 // run simulation in batch mode
 scicos_simulate(scs_m,list(),%scicos_context)
 // read back result in Scilab
 mopen("goo","rb")
 YT=mget(3*tf/dt+1000,"d")
 mclose()
 t=YT(3:3:$);u=YT(1:3:$);y=YT(2:3:$)
endfunction
```

The following script shows how the function syslinsim can be used:

```
// Example
Sys=(1-2*%s)/(1+%s+%s^2);
tf=20;dt=.01;
ui=0;
T=[ 2   4  7  9 ];
U=[ 2  -2  2  0 ];
[t,u,y]=syslinsim(Sys,tf,dt,ui,T,U);
// Display result
xset("window",1)
xbasc()
plot2d(t,u,2)
plot2d(t,y,5)
```

Figure generated by this script is given in Figure 11.4.

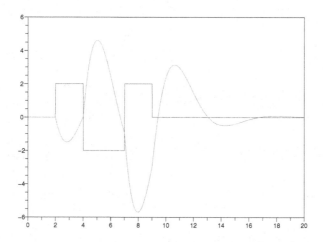

Figure 11.4. Simulation result displayed by Scilab.

11.2.3 Special blocks: From Workspace and To Workspace

Scicos provides two special blocks: From Workspace and To Workspace to facilitate the data exchange between Scilab and Scicos through files. By using these blocks, the exchanged data is seen on the Scilab side as Scilab variables and on the Scicos side as the output and input of two blocks. The use of files in this case is completely transparent to the user.

The Scilab variables associated with these blocks represent Scicos signals and are represented as Scilab structures with two fields: "time" and "values". The field "time" is a column vector containing time values in increasing order. The "values" field can be a matrix if the exchanged signal is a vector, each column of which giving the time history of an element of the vector.

The Scilab variable associated with the From Workspace block contains the sampled values of the Scicos signal at activation times of the block. The Scicos signal associated with the Scilab variable used in the To Workspace block is obtained by interpolation. Various interpolation strategies are available as block parameter.

The use of From Workspace and To Workspace blocks is the most convenient way of exchanging signals between Scilab and Scicos. For example its use allow us to modify the implementation of the previous example leading to a simpler batch processing setup. In particular, the Scicos diagram is modified as illustrate in Figure 11.5

The syslinsim function is modified as follows:

```
function [t,u,y]=syslinsimm(Sys,tf,dt,ui,T,U)
// Define symbolic parameters
%scicos_context.num=Sys.num
%scicos_context.den=Sys.den
%scicos_context.dt=dt
  // Scilab structure V is used to send data to Scicos
V.time=[0,T,tf+1]';
V.values=[ui,U,0]';
```

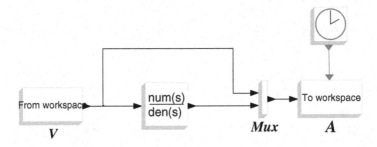

Figure 11.5. Using From Workspace and To Workspace blocks for exchanging data between Scicos and Scilab.

```
load("bgm.cos")
scs_m.props.tf=tf
// run simulation in batch mode
scicos_simulate(scs_m,list(),%scicos_context)
// read back result in Scilab
t=A.time;u=A.values(:,1);y=A.values(:,2);
endfunction
```

Often the purpose of batch processing Scicos simulations is to optimize system parameters. To illustrate how this can be done, consider the Scicos diagram in Figure 11.6.

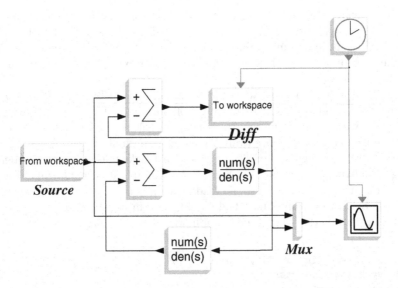

Figure 11.6. The From Workspace and To Workspace blocks used for parameter optimization.

The problem here is that of finding the optimal values of a and b, the PI controller parameters. We will use an optimization algorithm for that. For input, we use a square wave, and the system to control, used to test the program, is $1/(1+s+s^2)$.

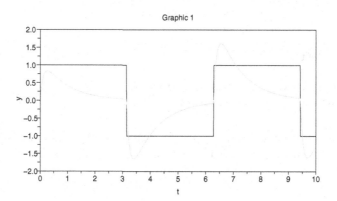

Figure 11.7. Simulation result without PI controller.

The square wave signal is generated by the **From Workspace** block based on the signal variable **Source** defined in Scilab, which includes the time instants of jumps in the signal and the associated values. The interpolation method is set to zero. This choice leads to a piecewise constant output, needed to generate a square wave signal.

Setting the PI controller parameters to zero $(a = b = 0)$, we obtain the simulation result in Figure 11.7. The optimization is then performed using the following Scilab script:

```
function e=eval_cost(p,z)
  %scicos_context.a=p(1)
  %scicos_context.b=p(2)
pause
  Info=scicos_simulate(scs_m,Info,%scicos_context)//,'nw')
pause
  e=Diff.values
  disp(norm(e))                          ← to monitor the progress of optimization
disp(p)
  endfunction

Source=struct();
Source.time=0;
Source.values=1;
load Untitled0m.cos
 for k=1:4
   Source.time(k+1)=k*%pi;
   Source.values(k+1)=(-1)^k;
 end

p0=[0;0];                                ← intial values of a and b

Info=list();
```

```
%scicos_context.a=p0(1);
%scicos_context.b=p0(2);

scicos_simulate(scs_m,Info,%scicos_context);
// datafit is an optimization function that
// uses optim but does not require the gradient
[p,err]=datafit(eval_cost,Z,p0,'ar',30);

%scicos_context.a=p(1);
%scicos_context.b=p(2);
// final simulation to visualize the solution
scicos_simulate(scs_m,list(),%scicos_context);
```

The function `datafit` reduces the distance between the input and output; the simulation result is depicted in Figure 11.8. The optimal values of the parameters are obtained by the program. `datafit` is discussed in Chapter 4.

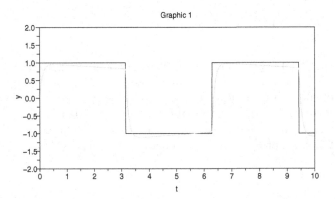

Figure 11.8. Simulation result with optimal PI controller.

The `From Workspace` block used to generate the desired input trajectory from within Scilab can be used to construct complex parameter optimization scenarios. For example, suppose the parameters a and b are to be adjusted not to just one input signal, but we want the PI controller to have good tracking capabilities for a step input *and* a ramp input. This is done in the following Scilab script:

```
function e=eval_cost(p,z)
  %scicos_context.a=p(1)
  %scicos_context.b=p(2)
  Source=struct();
  Source.time=z(1:2);Source.values=z(3:4);
  scicos_simulate(scs_m,Info,%scicos_context,'nw')
  e=Diff.values
  disp(norm(e))
endfunction

load Untitled0m2.cos
```

```
Z=[0 10 1 1;0 10 0 1/2]';

p0=[0.;0.];
Info=list();
%scicos_context.a=p0(1);
%scicos_context.b=p0(2);
 for z=Z
   Source=struct();
   Source.time=z(1:2);Source.values=z(3:4);
   scicos_simulate(scs_m,Info,%scicos_context);
 end
[p,err]=datafit(eval_cost,Z,p0,'ar',30);

Info=list();
 for z=Z
   Source=struct();
   Source.time=z(1:2);Source.values=z(3:4);
   %scicos_context.a=p(1);
   %scicos_context.b=p(2);
   scicos_simulate(scs_m,Info,%scicos_context);
 pause
 end
```

The interpolation method in this case is set to one (linear interpolation) in order to generate the ramp signal.

Note that the optimized cost in this case is the sum of the two distances obtained for a step and a ramp input. It is, of course, fairly straightforward to weigh the distances differently by using additional arguments in the datafit function. The simulation results without PI controller (parameters a and b set to zero) are given in Figure 11.9 and Figure 11.10. The simulation results using the optimal PI controller are given in Figure 11.11 and Figure 11.12.

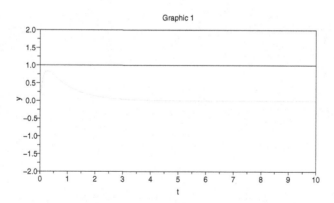

Figure 11.9. Simulation result for step input without PI controller.

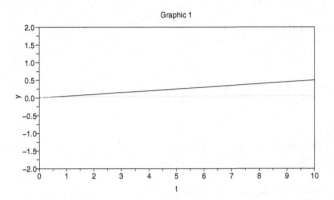

Figure 11.10. Simulation result for the ramp input without PI controller.

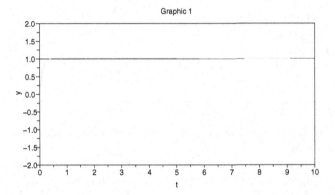

Figure 11.11. Simulation result for step input with optimal PI controller. The two curves are almost identical.

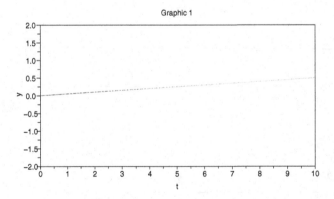

Figure 11.12. Simulation result for ramp input with optimal PI controller. The two curves coincide almost perfectly.

11.3 Steady-State Solution and Linearization

Scicos is primarily used to construct nonlinear models for simulation purposes. For analysis and controller or filter synthesis, it is often necessary to have a linearized version of the model, or more specifically, of the part of the model that represents the system being analyzed. Scicos provides some facilities for performing linearization of such systems.

The method presented in this section concerns only continuous-time components of the system described by a Scicos model or submodel (Super block). These components represent a finite-state, in general nonlinear, dynamical system that can be expressed as follows:

$$\dot{x} = f(x, u, t), \tag{11.1}$$

$$y = g(x, u, t). \tag{11.2}$$

Here x is the continuous-time state vector; u and y are the input and output vectors. For simplicity, we assume here after that f and g do not depend explicitly on time and drop t from their list of arguments.

An equilibrium point of system (11.1) is any pair (x_0, u_0) satisfying

$$f(x_0, u_0) = 0.$$

The state x_0 is the steady-state solution associated with the constant input u_0. This type of equilibrium is also sometimes called a setpoint. Clearly if the system is at an equilibrium point and u is the corresponding constant u_0, then the state remains at x_0 indefinitely.

In most control problems, the objective of the controller is to keep the system near an equilibrium point. Under mild regularity assumptions on the functions f and g, the behavior of the system near the equilibrium point can be approximated with a linear model.

Let

$$\delta x = x - x_0, \quad \delta u = u - u_0,$$

be the deviation from the equilibrium. Then the following linear model approximates the continuous-time behavior of the nonlinear system:

$$\dot{\delta x} = A\delta x + B\delta u,$$

$$\delta y = C\delta x + D\delta u,$$

where matrices A, B, C, and D are given by

$$A = \frac{\partial f}{\partial x}(x_0, u_0), \quad B = \frac{\partial f}{\partial u}(x_0, u_0),$$

$$C = \frac{\partial g}{\partial x}(x_0, u_0), \quad D = \frac{\partial g}{\partial u}(x_0, u_0).$$

To illustrate how linearization can be done and used in Scicos, we consider an example. In particular, we consider the classical problem of the control of an inverted pendulum on a cart. The system consists of a cart of mass M traveling on a ramp and subject to a force $u(t)$ with an inverted pendulum hinged on its side. For simplicity, all friction and the mass of the cart wheels are neglected. Moreover, it is assumed that the mass of the pendulum is concentrated at the tip and has value m. It is also assumed that the slope of the ramp, ϕ, is known. The setup is illustrated in Figure 11.13.

The dynamics of this mechanical system can be obtained easily from the Euler Lagrange equation, and is given below:

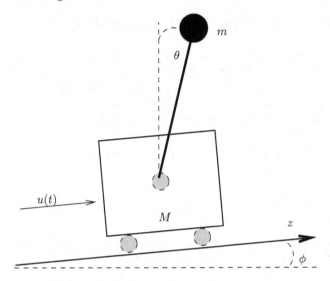

Figure 11.13. Inverted pendulum on a cart.

$$(M + m)\ddot{z} + ml\ddot{\theta}\cos(\theta - \phi) - ml\dot{\theta}^2\sin(\theta - \phi) = u(t) - (M + m)g\sin(\phi), \quad (11.3)$$
$$ml^2\ddot{\theta} + ml\ddot{z}\cos(\theta - \phi) - mgl\sin(\theta) = 0. \quad (11.4)$$

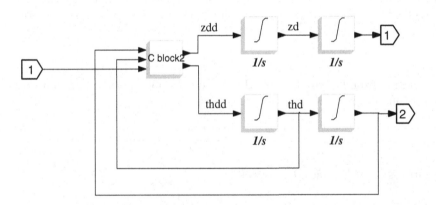

Figure 11.14. Scicos diagram for inverted pendulum on a cart problem.

To start, we need to construct a Scicos diagram to model these dynamics. We can do this in many different ways, both using existing blocks and constructing new blocks. To use existing blocks, we can reduce considerably the complexity of the diagram by using the `Mathematical Expression` block. The model can be reduced to a single block if a new block is used. This can be done with both C and Scilab blocks. Here we have chosen to use a new C block but leaving out the integrators; see Figure 11.14. The `Cblock4` is used to evaluate $\ddot{z}$ and $\ddot{\theta}$ as functions of θ, $\dot{\theta}$, and u. The expressions for $\ddot{z}$ and $\ddot{\theta}$ are readily obtained from (11.3) and (11.4), and are coded in the block as follows:

```
#include <math.h>
#include <scicos/scicos_block.h>
void inv_pend(scicos_block *block,int flag)
{
double M,m,l,g,ph,th,thd,zdd,thdd,u,delta;
M=block->rpar[0];
m=block->rpar[1];
l=block->rpar[2];
ph=block->rpar[3];
g=9.81;
th=block->inptr[0][0];
thd=block->inptr[1][0];
u=block->inptr[2][0];

delta=M*m*l*l+m*m*l*l*pow(sin(th-ph),2);
zdd=(m*l*l*(m*l*thd*thd*sin(th-ph)+u-
    (M+m)*g*sin(ph))-m*m*l*l*l*g*sin(th)*cos(th-ph))/delta;
thdd=(-m*l*cos(th-ph)*(m*l*thd*thd*sin(th-ph)+u-
    (M+m)*g*sin(ph))+(M+m)*m*g*l*sin(th))/delta;
block->outptr[0][0]=zdd;
block->outptr[1][0]=thdd;
}
```

Note that the block's parameter vector **rpar** contains M, m, l, and ϕ, which can be given numerical values or defined symbolically, for example as [M m l ph], in the dialog box of the block. In this latter case, Scilab variables M, m, l, and ph must have been previously defined in the context of the diagram.

The model non_lin can be used to construct a linearized model of the nonlinear system and subsequently a controller. It can also be used in the construction of a complete Scicos diagram for testing the controller.

11.3.1 Scilab Function steadycos

Computing an equilibrium point can be done using the Scilab function **steadycos**. The calling sequence is the following:

[X,U,Y,XP]=steadycos(scs_m,X,U,Y,Indx,Indu,Indy [,Indxp [,param]])

where

- scs_m: a Scicos diagram data structure.
- X: column vector. Continuous state. On input, it can be set to [] if zero.
- U: column vector. Input. On input, it can be set to [] if zero.
- Y: column vector. Output. On input, it can be set to [] if zero.
- Indx: indices of entries of X that are not fixed. If all can vary, must be set to 1:$.
- Indu: indices of entries of U that are not fixed. If all can vary, set to 1:$.
- Indy: indices of entries of Y that are not fixed. If all can vary, set to 1:$.
- Indxp: indices of entries of XP (derivative of x) that need not be zero. If all can vary, it must be set to 1:$. The default value is [].
- param: list with two elements (default list(1.d-6,0))
  - param(1): scalar. Perturbation level for linearization.
  - param(2): scalar. Time t.

The equilibrium point is, in general, not unique, and parts of the vectors x_0 and u_0 can be set a priori to desired values. Vector indices are used to specify parts of the state and the input that are free to be adjusted by the program. Even by specifying properly these indices, the equilibrium point may not be unique and the input arguments X and U can be used to direct the search toward a desired solution since they are used as initial guess values for the search algorithm.

Often the problem of finding an equilibrium point is that of finding a steady-state solution for a given input, in which case Indu is set to [] and Indx to 1:$. But in the example we are considering in this section, the equilibrium point we are looking for has an unknown u_0 and a known x_0. The desired objective is to keep the cart at the center with the pendulum pointing upward. The force u_0 is what is needed to prevent the cart from moving; that is, it compensates for the force of gravity. It is not difficult to see that

$$u_0 = (M + m)g\sin(\phi), \tag{11.5}$$

but we let the program find this by itself. This is done using the following commands:

```
⟼ load non_lin.cos
⟼ [X,U,Y,XP]=steadycos(scs_m,[],[],[],[],1,1:$)
 XP  =
    1.0E-16 *
!   0.3830958 !
!   0.         !
! - 0.3811819 !
!   0.         !
 Y  =
!   0. !
!   0. !
 U  =
    12.731756
 X  =
!   0. !
!   0. !
!   0. !
!   0. !
```

Here it is assumed that the Scicos diagram non_lin has been saved in the current directory.

Loading non_lin.cos provides the variable scs_m, which contains all the information regarding the diagram. This variable is then used as the argument to the function steadycos.

11.3.2 Scilab Function lincos

Computing the linearized system around an equilibrium point can be done using the Scilab function lincos. The calling sequence is the following:

```
sys= lincos(scs_m [,x0,u0 [,param] ])
```

where

- scs_m: a Scicos diagram data structure.
- x0: column vector. Continuous-time state around which linearization is to be done (default value is 0).
- u0: column vector. Input around which linearization is to be done (default is 0).

- param: list with two elements (default is list(1.d-6,0)).
  - param(1): scalar. Perturbation level for linearization.
  - param(2): scalar. Time t.
- sys: linear state-space system.

The linear system obtained from the linearization is coded in a "linear system" Scilab object. This is a tlist including the A, B, C, D matrices. Applied to our example, we obtain

```
⟼ sys= lincos(scs_m,X,U)
 sys  =
        sys(1)   (state-space system:)
!lss  A  B  C  D  X0  dt  !

        sys(2) = A matrix =
!   0.    0.   - 2.993E-08   - 2.9195682 !
!   1.    0.     0.            0.         !
!   0.    0.     2.978E-08    12.714983   !
!   0.    0.     1.            0.         !

        sys(3) = B matrix =
!   0.0997019 !
!   0.        !
! - 0.0992038 !
!   0.        !

        sys(4) = C matrix =
!   0.    1.    0.    0. !
!   0.    0.    0.    1. !

        sys(5) = D matrix =
!   0. !
!   0. !

        sys(6) = X0 (initial state) =
!   0. !
!   0. !
!   0. !
!   0. !

        sys(7) = Time domain =
  c
```

The linear system can now be used to construct a controller. Many methods are coded and available in Scilab for designing controllers including pole placement, LQG, H_∞, and LMI. Here we use a simple pole placement method:

```
Kc=-ppol(sys.A,sys.B,[-1,-1,-1,-1]);
Kf=-ppol(sys.A',sys.C',[-2,-2,-2,-2]);Kf=Kf';
Contr=obscont(sys,Kc,Kf);
```

The "linear system" Contr contains the system matrices of the controller.

To test and validate the controller's use in the original nonlinear model, a Scicos simulation must be performed. This is done using the Scicos diagram in Figure 11.15. The Super block non_lin contains the diagram in Figure 11.14. An animation block is used to show in real time the movement of the cart and the pendulum. This block is developed

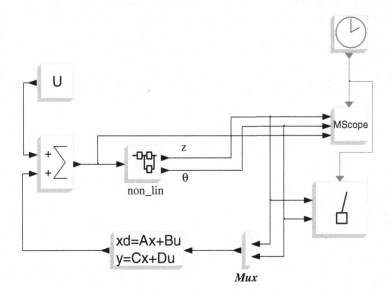

Figure 11.15. Complete Scicos diagram for testing the controller and validating the result by animation.

with a type-5 computational function. The interfacing and the computational functions are given in Section B.1.2.

Note that the control value (input to the Super block) is obtained by adding U, which is the value of u corresponding to the equilibrium point, denoted by u_0 earlier, and the output of the controller that generates δu.

To simplify the procedure for using this diagram, all the controller design steps are included in the context of the diagram as follows:

```
M=10;m=3;l=3;ph=0.1;
scs_m=scs_m.objs(27).model.rpar;
[X,U,Y,XP]=steadycos(scs_m,[],[],[],[],1,1:$);
sys= lincos(scs_m,X,U);
Kc=-ppol(sys.A,sys.B,[-1,-1,-1,-1]);
Kf=-ppol(sys.A',sys.C',[-2,-2,-2,-2]);Kf=Kf';
Contr=obscont(sys,Kc,Kf);
clear('scs_m','X','Y','XP','Kc','Kp','sys')
z0=-4;th0=.02;
```

Here we see that the scs_m corresponding to non_lin is extracted directly from the scs_m of the diagram, which is always available within the Scicos environment. For this, the object number of the Super block in scs_m must be known. In this case, it is the 27th. This information can be obtained using the Get Info button in the Object menu.

The context defines Scilab variables M, m, l, ph, z0, th0, U, and Contr, which are used as formal parameters of various blocks in the diagram. But there are intermediary variables such as Kc, Kf, which are not needed outside the context and are removed with the clear statement to avoid the needless storage of their values in the %scicos_context. The removal of scs_m is particularly important. If it is not removed, then it would clash with the main diagram's scs_m.

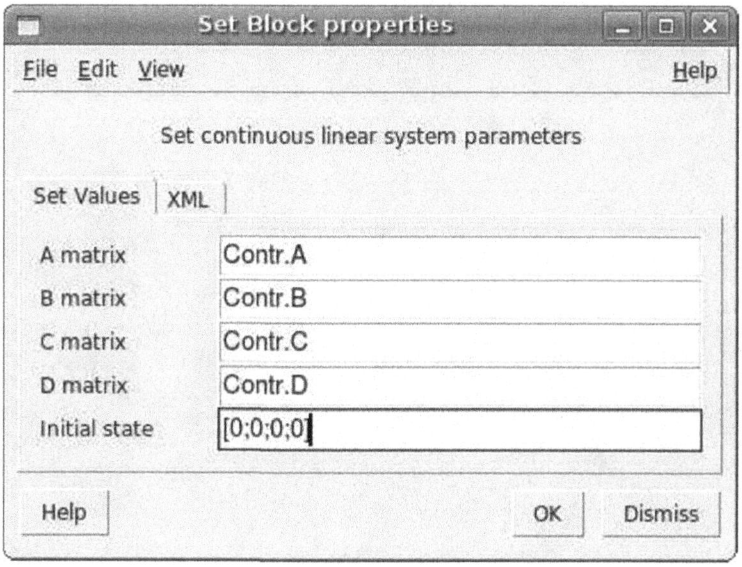

Set Block properties
File Edit View
Help
Set continuous linear system parameters
Set Values
XML
A matrix
Contr.A
B matrix
Contr.B
C matrix
Contr.C
D matrix
Contr.D
Initial state
[0;0;0;0]
Help
OK
Dismiss

Figure 11.16. Dialog box of the controller block.

To see how the Scilab variables defined in the context are used, consider for example the dialog box of the linear system block representing the controller illustrated in Figure 11.16. The variable `U` is used in the constant block feeding the summation block. The real parameters of the `CBlock4` block are set to `[M m l ph]`, and `z0` and `th0` are used as initial conditions for integration blocks in the `non_lin` diagram.

The simulation results confirm the correctness of the design procedure as shown in Figure 11.17. The system state and control converge to the desired steady-state values. The correctness of the model itself can be verified by examining the animation.

10
5
0
−5
−10
1.0
0.5
0.0
−0.5
−1.0
70
50
30
10
−10
0 3 6 9 12 15 18 21 24 27 30

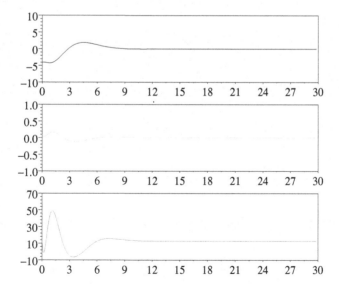

Figure 11.17. Simulation results. The displayed functions correspond to u, z, and θ.

12

Code Generation

Scicos provides a code-generation facility that can be used to create C code for a model defined in Scicos. Code generation has many applications. It can be used to improve simulation performance. Even though Scicos diagrams are compiled in the sense that all the scheduling tables are precomputed, there is still some overhead in using the Scicos simulator that is eliminated by running the code that has been generated. Even if the performance gain is not significant, replacing a Super block with a basic block may be of interest if the designer of the model does not want to reveal the content of the Super block. If the Super block is converted to a basic block, the designer can then simply distribute a compiled version of the generated code. But the most important application of code generation is the creation of standalone applications (programs that run independently of Scilab), which can be used, in particular, in real-time applications.

This chapter discusses how the code-generation facility can be used and what the limitations are. Some examples are provided.

12.1 Code Generation Procedure

The `Code Generation` item in the `Object` menu can be used to generate C code for the submodel defined within a Super block. In addition, once invoked, `Code Generation` compiles and links incrementally the generated code with Scilab and replaces the Super block with a basic block having for computational function the generated code. Except in some special situations that we shall discuss later, the replacement of the Super block with this new block should not alter simulation results in any way. `Code Generation` also creates a main C program implementing a standalone application.

Consider the Scicos diagram in Figure 12.1. In this model, a random number is generated by the `random generator` block at every clock event. The block parameter is set so that the random number can be both positive and negative. If the number is positive, the `If-then-else` block activates the memory block `1/z`, which is updated by adding its previous value to the (positive) value of the random number. Finally, the result is displayed in the scope alongside the values of the random signal.

The simulation result shows, as expected, that the output of the memory block increases only when the random number is positive. The output remains constant otherwise. See Figure 12.2.

Suppose now that we want to isolate the part of the diagram that receives a number and, depending on its sign, adds it to a counter, in order to perform code generation for

S.L. Campbell et al., *Modeling and Simulation in Scilab/Scicos with ScicosLab 4.4*,
DOI 10.1007/978-1-4419-5527-2_12, © Springer Science+Business Media, LLC 2010

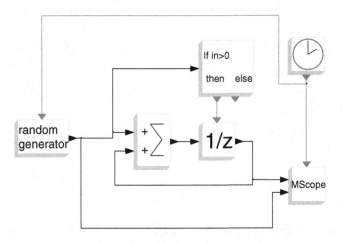

Figure 12.1. Original Scicos diagram.

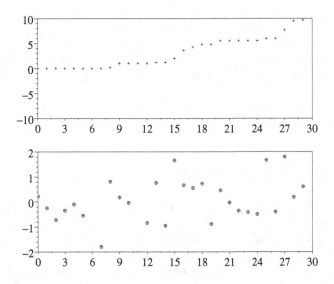

Figure 12.2. Simulation result for the Scicos diagram in Figure 12.1. Output is the top graph. The random numbers generated are shown in the bottom graph.

that part of the diagram. The first step is to place this part in a Super block. This can be done using the `Region to Super Block` facility. See Figure 12.3.

To generate C code for this Super block, it suffices to select the `Code Generation` item in the `Object` menu and then click on the Super block. A dialog box is then opened so that the user can specify the name of the new block and the directory in which the generated code should be placed. Once the answers are validated, the following lines are displayed in Scilab, indicating that the code has been successfully generated, compiled, and linked:

```
    building shared library (be patient)
shared archive loaded
```

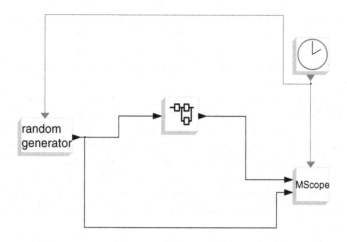

Figure 12.3. Scicos diagram in Figure 12.1 after isolating the part for which code generation is to be done.

```
Link done
Link done
```

Subsequently, the Super block is converted into a block; see Figure 12.4. Simulation can be performed to verify that this operation has not altered system behavior.

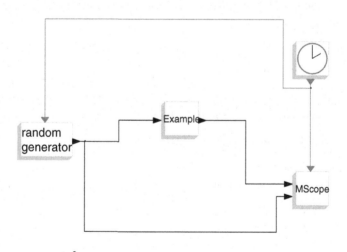

Figure 12.4. Scicos diagram after code generation.

This code generation operation creates a number of files in the specified directory. Since we have named our new block Example, all of the generated file names start with Example. The main file is Example.c, which is the computational function of the new block (this file is already compiled and linked with Scilab), and `Example_Cblocks.c`, `Example_standalone.c`, and `Example_act_sens_events.c`, which are used for standalone application. The file `Example_Cblocks.c` contains the main routines which are also found in `Example.c`. The file `Example_standalone.c` is the main standalone program.

The file `Example_act_sens_events.c` contains the routines corresponding to the input/output ports. In an embedded application, they would be the routines handling communications with sensors and actuators. That is why the routines are named `Example_sensor` for the inputs and `Example_actuator` for the outputs. The file also contains the routine `Example_events`, which is used to specify how the program should be executed by setting the event dates and ensuring real-time pacing if needed. This file is generated the first time code generation is performed, for a given example, and contains default routines, which consist of sensors and actuators that read from and write to the shell, and event dates regularly spaced in time with no real-time pacing. Since this file is very likely to be edited by the user in subsequent code generations for the same example, the user has the option of not regenerating it.

To create the standalone version, the following shell command should be used inside the directory where the files have been created: under Unix, Linux, and MacOSX

```
make -f Example_Makefile standalone
```

under Windows (inside an MSDOS shell)

```
nmake /f Example_Makefile standalone
```

Note that under Windows, Scilab `.dll` files must be accessible via the library path of the system, or should be copied into the current directory. These files are in the `SCI/bin` directory.

The compilation may generate some warnings concerning unused variables. This is normal. If everything goes well, this command creates the executable file **standalone** (**standalone.exe** under Windows). Running the program, for example, under Unix, Linux, and MacOSX, by typing

```
./standalone
```

the user is asked to enter an input value (followed by carriage return). In response, the program computes and displays the corresponding output value. Then the user is asked again to enter an input value, and so on. The following is what is displayed on the terminal. Note that the user has entered 23, 2 and 3; the rest is displayed by the program.

```
Require outputs of sensor number 1
time is: 0.100000
size of the sensor output is: 1
Please set the sensor output values
23
Actuator: time=0.100000, u(0) of actuator 1 is 0.000000
Require outputs of sensor number 1
time is: 0.200000
size of the sensor output is: 1
Please set the sensor output values
2
Actuator: time=0.200000, u(0) of actuator 1 is 23.000000
Require outputs of sensor number 1
time is: 0.300000
size of the sensor output is: 1
Please set the sensor output values
```

3
```
Actuator: time=0.300000, u(0) of actuator 1 is 25.000000
```

Of course this sequence continues until the final time `tf` is reached. Note that *sensor output* is the input to the program (block) and input (u) of the actuator is the output of the program.

These actuator sensor routines are, of course, useful only for debugging purposes. They should be adapted to the application at hand.

The standalone program can also be called with optional arguments:

```
standalone [-h] [-v] [-i arg] [-o arg] [-d arg] [-s arg] [-e arg] [-t arg]
```

where

- `-h`: prints a short help page,
- `-v`: prints the Scilab version used to create the program,
- `-i`: to define input file name; by default, terminal is used,
- `-o`: to define output file name; by default, terminal is used,
- `-d`: sets discrete-time clock period; default is 0.1,
- `-s`: to chose the fixed-step integration method among first-order Euler (1), Heun (2), and fourth order Runge Kutta (3),
- `-e`: sets discretization time period used by ode solver in case of continuous-time states, default is 0.001,
- `-t`: defines final simulation time, default is 30.

The options s and e concern only diagrams containing continuous-time states.

12.2 Limitations

We have seen that C code can be generated to realize the behavior of a Super block in Scicos simply by designating the Super block after selecting the `Code Generation` menu item. There are, however, limitations on the Scicos models that can be inside the Super block. This section presents these limitations.

12.2.1 Continuous-Time Activation

In most applications, even if the Scicos model contains continuous-time blocks, the part of the model for which code generation is to be performed contains only discrete-time components. Typically, this part represents a controller or a filter that is to be implemented in a real-time application on digital hardware. Scicos simulations in such cases are used to evaluate the controller/filter performance when applied to a continuous-time model of the environment.

In ScicosLab 4.4, the code generator has been extended and allows code generation for a much wider class of Super blocks, including continuous-time blocks with full support for zero-crossing events. In presence of continuous-time dynamics, the standalone code uses fixed-step solvers particularly appropriate for real-time applications.

12.2.2 Synchronicism

A new extension of Scicos allows code generation for Super blocks containing event generators. Synchronous events (events redirected by conditional blocks If-then-else and ESelect) were already allowed in earlier versions. But now code generation can also be applied to Super blocks containing SampleCLK blocks. The result is of course not just a basic block because, as we have already seen, a basic Scicos block cannot be self activated; its activation must come from the outside. The code generator in this case generates a new SampleCLK block in addition to a basic block. The new SampleCLK block is parameterized so that all the SampleCLK blocks in the original Super block are obtained from it by subsampling. The subsampling mechanism is of course included in the basic block.

A basic block, even combined with a SampleCLK block, cannot generate asynchronous events. This sets some limitations on the class of Super blocks that can be handled by Code Generation. Consider the Super block model depicted in Figure 12.5.

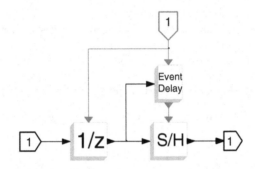

Figure 12.5. Asynchronous Super block diagram.

In this example, the 1/z and the S/H blocks are not synchronous. The reason is that S/H is activated by the output of the Event Delay block and the events corresponding to this output are not synchronous with the input events of this block that activate the 1/z block. In fact, no Scicos basic block can produce output events synchronous with its input events, the only exceptions being the If-then-else and event select block,s which, strictly speaking, are not blocks. So a hard constraint on the blocks within the Super block is that none should have an output activation port unless it is an If-then-else or an event select block.

12.3 A Look Inside

Code generation is done by the Scilab function CodeGeneration_, which can be found in the directory scilab/macros/scicos. This function takes a Scicos model (model inside the designated Super block) and converts it into a regular block by generating the corresponding computational function in the C language. It also generates a standalone C program. The code generation is done in five steps: diagram construction, compilation, code generation, block construction, and substitution.

Diagram Construction

The subdiagram within the designated Super block is extracted and used to construct a new Scicos diagram. This requires modifying the subdiagram because of the presence of input/output ports that are not Scicos blocks. These ports, used inside Super blocks, are eliminated at a precompilation stage when the original diagram is compiled.

The replacement of regular input ports is fairly straightforward because each input port is simply replaced with a regular block representing a sensor block and each output port is replaced with an actuator block. In a standalone application, it is up to the user to define the computational functions of these blocks. The Scicos code generator, by default, generates computational functions that read from and write to the terminal where the standalone application is launched. This facility is useful for debugging purposes.

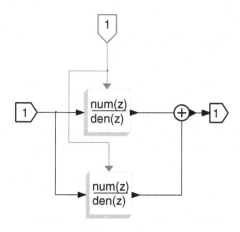

Figure 12.6. Diagram inside the Super block designated for code generation.

The activation of the sensor blocks must be done through the addition of activation links. Let us, consider the simple case illustrated in Figure 12.6. We see the diagram inside the Super block designated for code generation. To be able to use the Scicos compiler, this diagram is transformed into the one illustrated in Figure 12.7. This transformation is transparent to the user, who never sees this diagram.

We see that the sensor block replacing the input port has received an explicit activation from the block replacing the activation input port. The actuator block operates on inheritance and does not require explicit activation (placing an explicit activation link would not affect its behavior in this case).

If the Super block contains more than one input port, then each port is turned into a sensor block. And each sensor block is explicitly activated. But if the block contains more than one activation input port, the situation becomes more complicated because the events received through these ports may or may not be synchronous. For example, if the Super block contains two activation input ports, three events must be considered: event received on port 1, event received on port 2, synchronous events received on ports 1 and 2. That is why, in this example, the two activation input ports are replaced with a single block having 3 output activation ports. The activations are then used in such a way as to preserve the functionality of the Super block.

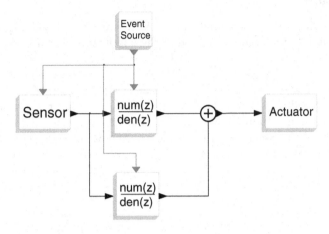

Figure 12.7. Diagram inside the Super block after transformation.

Compilation

The modified diagram corresponds to a valid Scicos diagram and it is compiled with the standard compiler. The result of the compilation contains all the scheduling tables needed to generate the simulation C code.

Code Generation

Two programs are generated: one to define a new block to substitute for the Super block in the Scicos diagram and one to be used for standalone applications. In the first case, the sensor and actuator blocks' computational functions are replaced with dummy programs. The inputs and outputs to the block are written directly on their outputs and inputs during simulation by the new computational function. For the standalone applications, default computational functions are generated in a separate file, which can be customized by the user.

Block Construction

The C code representing the computational function of the new block is generated at the "Code generation" stage. Here the Scilab interfacing function is generated.

Substitution

The new block replaces at this stage the Super block in the Scicos diagram. The compilation and linking of the associated computational function is done automatically so that after the substitution, the new Scicos diagram can be simulated.

12.4 Some Pitfalls

In most cases, code generation does not alter the behavior of the system. The Super block and the block replacing it have identical behavior. There is, however, one situation in

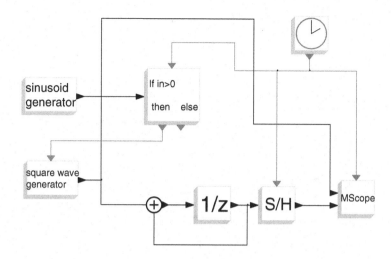

Figure 12.8. Scicos diagram posing problems for the code generator.

which the two behaviors may differ. This difference is due to the way inheritance works in Scicos. Consider the Scicos diagram illustrated in Figure 12.8.

In this diagram, we use the 1/z block, available in the **Linear** palette, with the inheritance option. The behavior of this block is exactly the same as the usual 1/z except that since it does not have an activation input port, it inherits its activation. Even though it is not recommended in Scicos to use blocks with discrete states that inherit their activations, there is nothing in the formalism that forbids it, and in some cases, it could be a useful facility.

Suppose now that we want to generate code for the part of the diagram including this new block, the **Summation** and the **S/H** (sample and hold) block. This is done by encapsulating the region of interest in a Super block using the **Region-to-Superblock** facility. The result is given in Figure 12.9.

By examining Figure 12.8, we see that the 1/z and **Summation**, which form a discrete integrator, are activated by the output of the **If-then-else** block, that is, by a subsampled version of the activations generated by the **Event Clock**. In fact, the activation takes place only if $\sin(t)$ is positive. On the other hand, the **S/H** block is activated directly by the **Event Clock**. Since this block simply performs a copy of the input on the output, its more frequent activation does not change the result of the simulation in any way. The encapsulation process does not alter, of course, the simulation result either, and for both Diagrams 12.8 and 12.9, the simulation result is given in Figure 12.10.

After code generation, we obtain the diagram in Figure 12.11, which results in the simulation given in Figure 12.12. Clearly the simulation results are different from those in Figure 12.10.

To explain the discrepancy, note that after the encapsulation step (Figure 12.9), the Super block contains both blocks activated through inheritance and blocks activated directly through the activation signal received on its activation input port. But when the Super block is converted into a basic block by code generation, the new block cannot exhibit both behaviors. It can work either by inheritance or by explicit activation, but not both. In this case, since the Super block has an explicit activation port, it has no

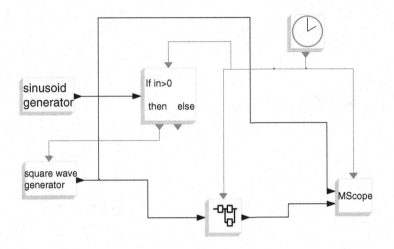

Figure 12.9. Diagram after encapsulation by Super block.

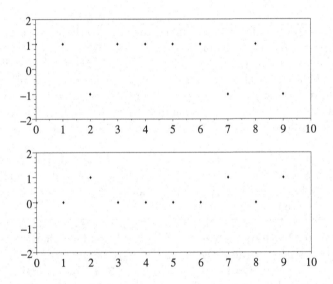

Figure 12.10. Simulation result before code generation (the output of the MScope display).

choice but to consider this port as the source of activation for all blocks inheriting their activations with the Super block. In a sense, the Super block is considered a functional encapsulation by the code generator and not just a graphical commodity.

Another situation in which such a problem occurs is when a block inherits from the new generated block. Consider the Scicos diagram in Figure 12.13. In this diagram, the scope inherits its activation from the S/H block, which means that it displays the sample values of $\sin(t)$ only if this latter is positive, as can be seen from the simulation result illustrated in Figure 12.14.

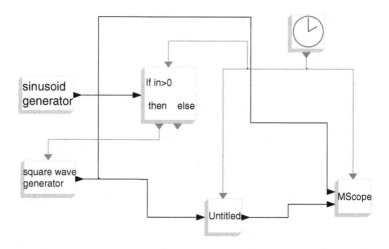

Figure 12.11. Diagram after the code generation step.

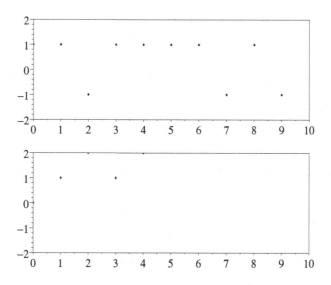

Figure 12.12. Simulation result after code generation. The output of the Untitled block diverges.

If the subsampling part of the diagram is converted into a block using code generation as illustrated in Figure 12.15, the simulation result changes because now the scope inherits from the new block and thus it is activated every time. The simulation result is illustrated in Figure 12.16.

12.5 Applications

Hardware in the loop and real-time embedded control are the main applications of code generation in Scicos. Hardware in the loop applications can be developed by constructing

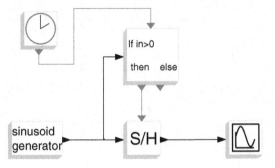

Figure 12.13. Original diagram; the scope inherits from the S/H block.

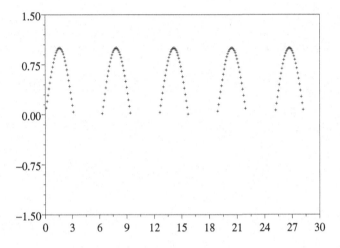

Figure 12.14. Simulation result shows that scope is active only if the value to be displayed is positive.

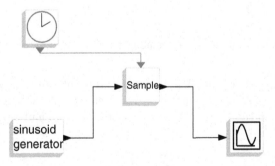

Figure 12.15. Diagram after code generation; the scope inherits from the Sample block.

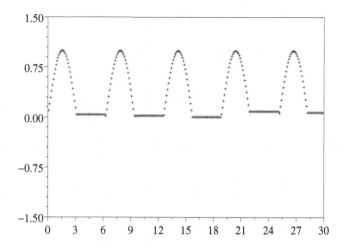

Figure 12.16. Simulation result is different from the original simulation, as expected.

specific Scicos blocks for interfacing control and measurement devices. Under Linux, for example, this has been done using the COMEDI tool (`http://www.comedi.org`). The Scicos COMEDI toolbox can be downloaded from the contribution section of the Scilab home page.

Direct real-time code generation for Linux RTAI is available through the RTAI-Lab project; see `http://a.die.supsi.ch/~bucher/rtai.html`. An adapted version of the Scicos code generator is used in this project. RTAI-Lab relies on Scicos for the control system code design and generation, with only the addition of some specific blocks and building options. The generated code is then embedded in an RTAI framework to be executed in soft/hard real time and monitored by a generic graphical user interface.

13

Debugging

When a large Scicos diagram does not yield the expected simulation results, it can be fairly complicated to identify the problem. To help the user, debugging tools are provided in Scicos. We shall present these tools later in this chapter, but we will first explain Scicos error messages.

13.1 Error Messages

When something goes wrong during the simulation, the simulation stops and an error message is displayed. Problems encountered by the simulator can be of a number of different types.

13.1.1 Block Errors

An error could occur within a block, or more specifically the computational function of the block. These errors must be signaled by the block itself using the set_block_error function, which should be called with a negative value depending on the type of error encountered. The following error flag values can be used:

- -1: This means that a block has been called with input out of its domain. In general, this means that one of the input arguments of the computational function does not have the expected value. The block can print out, in a Scilab menu, more details using the scifunc function.
- -2: This error is signaled if the block encounters a singularity, that is, an operation such as division by zero.
- -16: This error corresponds to the case that a block wants to allocate memory (for which it has to use the scicos_malloc function) but it cannot.

Another block-related error, not detected by the block itself but by the simulator, happens when the block tries to program an event on an output activation port but the event previously programmed on the same activation port has not been fired. This happens, for example, if an event delay block is activated with an event clock and the period of the clock is smaller than the delay period in the delay block.

Finally, it could happen that Scicos simply cannot find the block's computational function. This could happen with a user block if the computational function is not linked in Scilab before simulation or its name does not match the name in the model data structure of the block (set by the interfacing function). The message in this case is unknown block.

S.L. Campbell et al., *Modeling and Simulation in Scilab/Scicos with ScicosLab 4.4*,
DOI 10.1007/978-1-4419-5527-2_13, © Springer Science+Business Media, LLC 2010

13.1.2 Errors During Numerical Integration

Scicos uses LSODAR in the explicit case to simulate the continuous-time dynamics of the model. When an error is signaled by lsodat, Scicos displays the error message integration problem and specifies the value of the error flag istate returned by LSODAR. The following values of istate are relevant in Scicos:

- -1: This means that too many calls have been made by the solver over the integration period. This often happens when the system is very stiff or nonsmooth and/or the integration period is too long. The integration period is the period between two event,s which often correspond to the activations generated by an event clock to drive a scope or read-from/write-to file blocks. In that case, to reduce the integration period it suffices to reduce the period of the event clock. It is also possible to set a maximum for the integration period in the Setup menu.
- -2: This means that excess accuracy has been requested (tolerances are too small). The tolerances can be set by the Setup menu.
- -5: This means that repeated convergence failures have occurred, probably due to the requested tolerances being too small.

In the implicit case, the solver DASKR can return additional error messages:

- -7: The nonlinear system solver in the time integration could not converge.
- -8: The matrix of partial derivatives appears to be singular.
- -9: The nonlinear system solver in the integration failed to achieve convergence, and there were repeated error test failures in this step.
- -12: Failure to compute the initial state $(x, \dot{x})$.

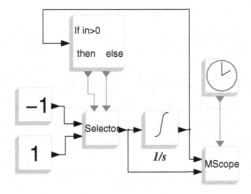

Figure 13.1. A simple system exhibiting sliding mode behavior.

Another error condition related to continuous-time dynamics, which is not always detected by Scicos, is the presence of a sliding mode. When a model exhibits sliding mode behavior, the modes of one or more blocks cannot be set properly and the simulation may cease to advance. Consider, for example, the system $\dot{x} = -1$ if $x \geq 0$ and $\dot{x} = 1$ if $x < 0$ modeled in Diagram 13.1. For this system the dynamics on either side of the zero-crossing surface move toward the surface. If a zero-crossing is used in the If-then-else block, then simulation halts as soon as the state of the integrator reaches zero. If zero-crossing (and thus mode) is not used, the simulation may proceed in some cases, but in most cases the numerical solver returns an error. For well-chosen simulation parameters, simulation can proceed and the result is given in Figure 13.2. This is a *reasonable* result from a practical point of view, but such sliding mode conditions usually imply a modeling error and should be avoided.

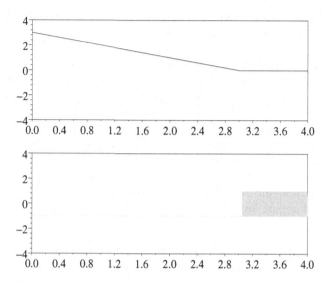

Figure 13.2. Simulation result without zero-crossing in the `If-then-else` block. The simulation gets stuck at $t = 3$ if zero-crossing is used. Top graph is the state. Bottom graph is $\dot{x}$.

13.1.3 Other Errors

Initial Conditions not Converging

At the start of the simulation, after all the blocks have been called with flag 4, Scicos calls all the blocks with flag 6 until the states and output values reach a stationary point. This is done by performing a fixed-point iteration, and the error is declared if this iteration does not converge. Few blocks need to use specifically flag 6.

Cannot Allocate Memory

The Scicos simulator needs to perform dynamic memory allocation when it is called. This error is encountered when the operating system cannot furnish the requested memory. This error is extremely unlikely on modern computers even for very large diagrams.

13.2 Debugging Tools

Information can be displayed during Scicos simulation on the Scilab window. How much information is displayed depends on the debugging level. At level 1, only event number at activation times and calls to the ODE or DAE solver are displayed. At level 2, in addition to the information displayed at level 1, the block number and activation time of every activated block is displayed.

The debugging level can be set to i by the Scilab command

```
scicos_debug(i)
```

It can also be set by a Scicos menu. If i is zero, the debugging is turned off.

In most cases, the display of information is not enough to efficiently debug Scicos diagrams. That is why Scicos provides a powerful debugging facility through the Debug block (in the Others palette). This block should be placed in the diagram being debugged.

The Debug block has a Scilab computational function of type 99. This type of block is not activated itself but it impersonates all the other blocks in the diagram, once before and once after their activations, provided the debug level is set to 2. The debug level is automatically set to 2 when the dialog box of the Debug block is opened and validated.

In the dialog box of the Debug block, it is possible to place Scilab instructions to be executed when the computational function of the Debug block is called. This means once before and once after each block activation. The default instruction in the dialog box of the Debug block is pause. This corresponds to a simulation in single-step mode. Under pause, all the block parameters can be inspected and even modified. To stop the debugging, it suffices to set the debugging level back to zero as follows:

```
-1-> scicos_debug(0)
```

Having the possibility to use any Scilab instructions provides a powerful debugging environment. For example, the pause statement can be made conditional:

```
if flag==1&scicos_time()>3&curblock()==2&block.inptr(1)>0 then
  pause
end
```

This way, the user can choose exactly where and when the pause should occur. But many other Scilab instructions can be used, as we shall see in the examples later. In these instructions, we have access to the environment available to any Scilab computational function of type 5. We have seen in the above example the use of functions scicos_time and curblock. The other functions available to type 5 blocks presented on page 215 can also be used, in addition to the function scicos_debug_count, which returns the number of times the block Debug has been called.

In addition to these functions, which provide information about the block, the Scilab function getscicosvars can be used to obtain information about the global state of the model, in particular x, z, and outtb. The continuous-time state x is obtained by the command getscicosvars(``x''), etc.

Other information that can be obtained from getscicosvars is xptr, zptr, rpar, rpptr, ipar, ipptr, inpptr, outptr, inplnk, outlnk, and lnkptr. These variables are part of %cpr.sim and do not evolve as a function of time. They are essential in determining which parts of x, z, and outtb correspond to which block. For more details, see Appendix A.

13.3 Examples

The Debug block can be used in various ways to supervise, analyze, and debug a Scicos diagram. Here we present two elementary examples to illustrate the possibilities. Typical applications are, of course, in general a lot more complex.

13.3.1 Log File

The Debug block can be used to create a log file including information concerning every single call to every block. The information recorded in the log file can be more or less detailed. Here we record the time, the block number, the values of inputs, and the computed values of the outputs. We do not consider the states or the way they are updated. It is for this reason that we consider only the case of flag equal to 1 in simulation phase 1. The following statements are placed inside the Debug block:

```
if flag==1&phase_simulation()==1 then
unit=mopen('scicoslog.dat','a')
if scicos_debug_count()==1 then mfprintf(unit,'SIMULATION DATE '+date()+'\n'),end
if scicos_debug_count()-int(scicos_debug_count()/2)*2>0 then
  mfprintf(unit,'Block number: %3d at time %5.3f \n',curblock(),scicos_time())
  for i=1:size(block.inptr)
    IN=strcat(string(block.inptr(i)'),';')
    mfprintf(unit,'   Input number '+string(i)+' is ['+IN+']\n')
  end
else
  for i=1:size(block.outptr)
    OUT=strcat(string(block.outptr(i)'),';')
    mfprintf(unit,'   Output number '+string(i)+' is ['+OUT+']\n')
  end
end
mclose(unit)
end
```

Note that the information concerning the inputs is written before the block is called, and the information concerning the outputs is written after the block is called. This is done by testing whether or not scicos_debug_count is odd. Since the Debug block is called twice for each block execution (once before and once after), if scicos_debug_count is odd, then it corresponds to a call before block activation.

The information, at each call, is appended to the end of the file scicoslog.dat. This file typically contains

```
SIMULATION DATE 11-Jan-2005
Block number:   1 at time 0.000
   Input number 1 is [-0.0761649]
   Output number 1 is [-2;1;2]
Block number:   3 at time 0.000
   Input number 1 is [0]
   Input number 2 is [-5.1904856;-2.9075105;4.7414013]
   Output number 1 is [0;0;0]
Block number:   8 at time 0.000
   Output number 1 is [0]
Block number:   2 at time 0.000
   Input number 1 is [-2;1;2]
   Output number 1 is [3;8]
```

and so on.

13.3.2 Animation

Here we use the Debug block to animate the block executions on the Scicos editor. This is done by highlighting a block when it is activated.

```
xset('window',1000) // Main Scicos editor window
k=%cpr.corinv(curblock())
if scicos_debug_count()-int(scicos_debug_count()/2)*2>0 then
    hilite_obj(k(1))
else
    xpause(500000)
    unhilite_obj(k(1))
end
```

Note that k gives the path to the block to be highlighted, but if the block is inside a Super block, only the first value of k is used. This means that only the Super block on the main window containing the block is highlighted.

The xpause command is used to make sure that the period of time when the block remains highlighted is sufficiently long so that it can be seen on screen.

14

Implicit Scicos and Modelica

14.1 Introduction

Standard Scicos is not well suited for physical component-level modeling. For example, when modeling an electrical circuit, it is not possible to construct a Scicos diagram with a one-to-one correspondence between the electrical components (resistor, diode, capacitor,...) and the blocks in the Scicos diagram. In fact, the Scicos diagram does not look anything like the original electrical circuit.

In this chapter we will discuss extensions to Scicos that make physical component-level modeling possible. But first we will take a simple linear circuit and model it in standard Scicos for comparison purposes. Consider the electrical circuit depicted in Figure 14.1. This circuit contains a voltage source, a resistor, and a capacitor.

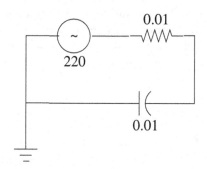

Figure 14.1. A simple electrical circuit.

To model and simulate this diagram in Scicos, we can use Kirchoff's law, which states that the voltage drops (or increases) around a circuit loop must add up to zero. This gives us

$$V_s + Ri + \frac{1}{C} \int i \, dt = 0, \tag{14.1}$$

where V_s is the voltage across the voltage source and i the current through the circuit. If we are interested in the voltage V across the capacitor, we can compute it as follows:

$$V = -\frac{1}{C} \int i \, dt. \tag{14.2}$$

S.L. Campbell et al., *Modeling and Simulation in Scilab/Scicos with ScicosLab 4.4*,
DOI 10.1007/978-1-4419-5527-2_14, © Springer Science+Business Media, LLC 2010

We can now implement these two equations in Scicos by using an integrator block receiving i as an input. The resulting Scicos diagram is depicted in Figure 14.2. Note that this Scicos diagram does not look anything like the original electrical circuit in Figure 14.1.

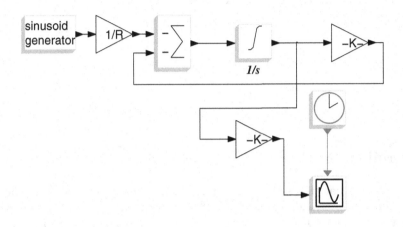

Figure 14.2. Scicos diagram realizing the dynamics of the electrical circuit in Figure 14.1.

An extension of Scicos allows us to model physical components directly within the standard Scicos diagrams. This is done by lifting the causality constraint on Scicos blocks and by introducing the possibility of describing block behaviors in the Modelica language. With this extension we can model naturally not only systems containing electrical components but also mechanical, hydraulic, thermal, and other systems. For example, the electrical circuit in Figure 14.1 can be modeled and simulated by constructing the Scicos diagram in Figure 14.3. The electrical components come from the **Electrical** palette.

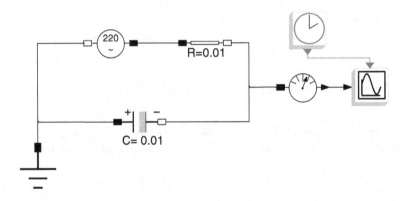

Figure 14.3. Scicos diagram using Modelica components.

This extension is still being developed. In this chapter we provide some information on how to use this facility, but we do not provide detailed documentation because things may change in the future. We also assume that the reader is familiar with Modelica; if not, consult www.modelica.org for documentation. More information on challenges posed by the Modelica/Scicos integration can be found in the following publications: [50], [51], [47], [46], [49].

14.2 Internally Implicit Blocks

The first step in extending Scicos to accept implicit blocks has been the introduction of *internally implicit* blocks. These blocks are identical to standard Scicos blocks except for the fact that the internal continuous-time dynamics are allowed to be implicit. This means that (9.6) for internally implicit blocks changes to

$$e(t) = f_0(t, x(t), \dot{x}(t), z(t), u(t), n_{\text{evprt}}, \mu(t)), \qquad (14.3)$$

where $e(t)$ is the residual, which in theory equals zero, and the solver tries to keep it close to zero during numerical integration.

Internally implicit blocks have explicit input-output ports and are used in Scicos exactly the same way as standard blocks are. This means, in particular, that the editor and, except for some minor modifications, the compiler, were not affected by the introduction of these blocks. But the simulator had to be extended because a diagram containing even a single internally implicit block results in a system of differential algebraic equations (DAE) that cannot be solved with an ODE solver. It is for this reason that the DAE solver IDA from the SUNDIALS suite had been interfaced with the Scicos simulator. DAEs were discussed earlier, in Chapter 3 where the solver dassl was presented. IDA is a modification of DASPK which is also a BDF based code like dassl is. Note that a different variant of DASPK called DASKR which also has root finding capability was used in earlier versions of Scicos. IDA is not accessible directly from Scilab. Users interested in using IDA or the ODE code CVODE will need to do so through Scicos.

It is important to note that the root finding part of IDA has been modified. The original version of IDA assumed that any switching surfaces were transversed and failure of this happening led to integration termination. However, in hybrid systems it is not unusual to have systems move along the switching surface after contact has been made. The IDA root finding routines have been modified in Scicos to allow for moving along the switching surface after contact is made with the surface if that is what is indicated by the model equations.

Solving DAE systems poses specific problems [12], in particular, that of finding consistent initial conditions as pointed out in Sections 3.1.4 and 3.2.4. In the case of a DAE, and in particular an internally implicit block, the state is not just x but the pair $(x, \dot{x})$, and not all pairs $(x, \dot{x})$ are consistent with the system equations. Finding consistent initial conditions can be a complex problem in some cases. This is particularly difficult in the multimode case, which is often encountered in the Scicos environment. To help the DAE solver find consistent initial conditions, an internally implicit block must be able to furnish information about the nature of each state variable. In particular, it should specify which state variables are algebraic (their derivatives do not appear in the equations) and which ones are differential (their derivatives do appear in the equations) [15]. This is done by the computational function of the block when it is called with flag 7.

We do not give detailed information here on how to construct internally implicit blocks. It suffices to say that the computational function is of type 10005, that $(x, \dot{x})$ must be placed in the continuous-time state slot in the model, that the computational function must return the residual $e(t)$ when it is called with flag 0 and should provide additional information when called with flag 7.

The reason we do not insist on the details is that one rarely needs to construct such an internally implicit block. These blocks are generated automatically when the diagram contains implicit blocks, as will be discussed in the next section.

14.3 Implicit Blocks

Implicit blocks, unlike internally implicit blocks, have implicit ports. An implicit port is different from an input or an output port in that connecting two such ports imposes a constraint on the values at these ports but does not imply the transfer of information in an a priori known direction, as is the case when we connect an output port to an input port. For example, the implicit block `Capacitor` used in the diagram of Figure 14.3 has current and voltage values on its ports but there is no way a priori, without analyzing the full diagram, to designate the ports as input or output.

Implicit blocks and the construction of diagrams based on these blocks is fully in the spirit of the object-oriented Modelica language [26]. The description of block behavior in Scicos is done in Modelica.[1]

14.3.1 Scicos Editor

Implicit and regular Scicos blocks can be used in the same diagram. The way implicit blocks are handled by the Scicos editor is similar to the way regular blocks are handled. Blocks can be copied from palettes or inserted using the `Add New Block` functionality. Two elementary palettes are provided by Scicos. Super blocks can be used to structure diagrams with implicit blocks. Note that Super block implicit ports are available in the `Sources` and `Sinks` palettes for that. A Super block can have both explicit and implicit ports.

Links are constructed as in the regular case. It is, of course, not possible to connect an implicit port to an explicit port (it would be meaningless anyway). The connection to the explicit world can come only from the implicit blocks having one or more explicit ports such as the voltage sensor in Diagram 14.3.

14.3.2 Scicos Compiler

The compilation with implicit blocks is again transparent to the user. During this compilation, the compiler performs a first stage compilation by grouping all the implicit blocks into a single internally implicit block. This is done by generating a Modelica program for the implicit part of the diagram and generating the computational function of the new internally implicit block. The C program is compiled (this requires a C compiler) and linked with Scilab, and the new internally implicit block replaces the implicit part of the diagram. The last stages are similar to the way code generation works, as we have seen in Chapter 12, except that the user does not see the implicit blocks being replaced.

After this first compilation stage, the compilation proceeds as with standard Scicos diagrams, and subsequently, so does the simulation.

---

[1] For the moment only a subset of the language is implemented.

14.3.3 Block Construction

As in the case of standard blocks, we need to provide two functions for each new block: an interfacing function, in Scilab, and a computational function, in Modelica. We can avoid providing an interfacing function, as we did in the case of regular blocks using the `CBlock4` block, by providing the Modelica code interactively in the `MBlock` block. We will discuss `MBlock` later in Section 14.5.

The interfacing function is similar to that of a regular block. The only difference is an additional field in the `model` structure and two additional fields in the `graphics` structure of the block.

`model.equations` is in fact present in all blocks. But in regular blocks, it is set to an empty `list` and not used. In the `VoltageSensor` block, for example, `model.equations` is defined as follows:

```
mo=modelica()
mo.model='VoltageSensor'
mo.inputs='p';
mo.outputs=['n';'v']
model.equations=mo
```

Implicit blocks can be classified either as inputs or outputs. Here the Modelica variable in the Modelica model `VoltageSensor` p is defined as input and n and v as outputs. But in fact, p and n correspond to implicit ports. The output v is explicit. It measures the voltage across the block, i.e., between p and n.

The implicit/explicit nature of the input-output ports is characterized in two new fields in the `graphics` structure. In the case of `VoltageSensor`, we have

```
graphics.in_implicit=['I']
graphics.out_implicit=['I';'E']
```

The interfacing function of this block, as with the rest of the blocks in the `Electrical` palette, can be found in the directory `scilab/macros/scicos_blocks/Electrical`. This directory contains also the Modelica sources of the blocks.

The Modelica program defining the behavior of the `VoltageSensor` block can be found in the file `VoltageSensor.mo`:

```
model VoltageSensor
  Pin p;
  Pin n;
  Real v;
equation
    p.i = 0;
    n.i = 0;
    v = p.v - n.v;
end VoltageSensor;
```

For a Modelica model to be usable in the definition of an implicit block in Scicos, the corresponding .mo file must be precompiled using the Modelica compiler `modelicac`, available in the directory `scilab/bin/modelicac`. The result is a file with extension .moc. For example, if the model is named `foo` and is defined in the file `foo.mo`, the command

```
modelicac -c foo.mo
```

produces a file named `foo.moc`. To use `foo` in Scicos, the Scilab variable `modelica_libs` must include the path to the directory where `foo.moc` can be found. Initially, the paths to the directories `Electrical` and `Hydraulics` are placed in `modelica_libs`. Paths to user-defined *.moc files must be added by the user to `modelica_libs` before launching Scicos or by activating the Scilab window.

External Function Calls

In Modelica, C and fortran external functions can be used. Scicos supports external function calls used in Modelica. We illustrate the use of external functions using an academic example. We consider in particular a voltage source where the output voltage is defined by an external C function:

```
model VsourceVar
  Pin p, n;
  Real v;
  parameter Real V0 "Amplitude";
equation
  V0 * Myfunc(time) = v;
  v = p.v - n.v;
  0 = p.i + n.i;
end VsourceVar;
```

Myfunc is not a Modelica built-in function; it must be defined as an external function as follows:

```
function Myfunc
  input Real u;
  output Real y;

  External "c";
end Myfunc;
```

The C code of the `Myfunc` function is defined in the include file `Myfunc.h`:

```
#include <math.h>
float Myfunc (float);
```

and in `Myfunc.c`:

```
#include "Myfunc.h"
float Myfunc(float u)
{
  float y;
  y = sin(u);
  return y;
}
```

This external function is given as an example; it simply realizes the sine function. We could obtain the same result simply by using Modelica's sine function.

To use this model in Scicos, we need to generate a shared library (`.dll` under Windows or `.so` under Linux and Unix) by compiling the C function and placing it in a directory along with all the Modelica files (`.mo` and the include file `Myfunc.h`. The path of the directory must be added to `modelica_libs` variable. Suppose the directory is called `mylibs`, then in Scilab this can be done by executing the following instruction:

modelica_libs=[modelica_libs,"mylibs"]

14.4 Example

Here we consider the construction and the use of a new implicit block in Scicos. This block is an ideal operational amplifier (infinite gain) to be used with the electrical components already available in the `Electrical` palette. The Modelica model of this operational amplifier is very simple and is taken from the Modelica library `Electrical.Analog.Ideal`, with small modifications to make it compatible with `modelicac`:

```
class IdealOpAmp3Pin
  Pin in_p "Positive pin of the input port";
  Pin in_n "Negative pin of the input port";
  Pin out "Output pin";
equation
  in_p.v = in_n.v;
  in_p.i = 0;
  in_n.i = 0;
end IdealOpAmp3Pin;
```

The Modelica program indicates that there are no currents flowing through the pins in_p and in_n, and that their voltages are equal. The pin out, on the other hand, can have any voltage and produce any current.

The interfacing function is defined as follows:

```
function [x,y,typ]=IdealOpAmp3Pin(job,arg1,arg2)
  x=[];y=[];typ=[]
  select job
    case 'plot' then
    standard_draw(arg1)
    case 'getinputs' then
    [x,y,typ]=standard_inputs(arg1)
    case 'getoutputs' then
    [x,y,typ]=standard_outputs(arg1)
    case 'getorigin' then
    [x,y]=standard_origin(arg1)
    case 'set' then
    x=arg1;
    case 'define' then
  model=scicos_model()
  model.in=[1;1];model.out=1;
  model.sim='IdealOpAmp3Pin'
  model.blocktype='c'
  model.dep_ut=[%t %f]
  mo=modelica()
  mo.model='IdealOpAmp3Pin'
  mo.inputs=['in_p';'in_n'];
  mo.outputs=['out']
  model.equations=mo
  exprs=[string([1])]
  gr_i=['txt=[''OpAmp''];';
        'xstringb(orig(1),orig(2),txt,sz(1),sz(2),''fill'')']
  x=standard_define([2 2],model,exprs,gr_i)
  x.graphics.in_implicit=['I';'I']
  x.graphics.out_implicit=['I']
  end
```

endfunction

Note that this block has no parameter (thus no dialog box) and all of its ports are implicit.

To use this block, we first construct the file `IdealOpAmp3Pin.moc` using the shell command

```
modelicac -c IdealOpAmp3Pin.mo
```

and in Scilab, we add the directory where this file resides to the Scilab variable `modelica_libs`. Note that the shell command can be executed from within Scilab using the **host** command:

```
host(SCI+'/bin/modelicac -c IdealOpAmp3Pin.mo')
```

Finally we load the interfacing function in Scilab:

```
getf('IdealOpAmp3Pin.sci')
```

The block `IdealOpAmp3Pin` is now available in Scicos and can be placed in a block or a palette using the **AddNewBlock** facility. This is done for the Scicos diagram in Figure 14.4. We have used the block `IdealOpAmp3Pin` to implement a voltage amplifier. The simulation result is given in Figure 14.5.

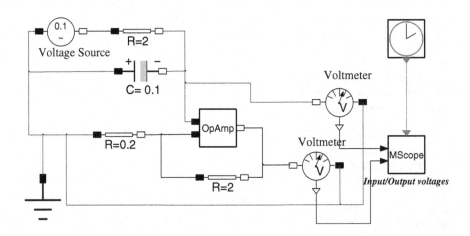

Figure 14.4. Scicos diagram containing the new operational amplifier block.

14.5 Scicos Block `MBlock`

The Scicos block `MBlock` (Modelica generic) allows us to define an implicit block interactively by providing the associated Modelica code without having to develop an interfacing function. In many ways, `MBlock` resembles the `CBlock4` block, except that the Modelica language replaces C for defining block's simulation function.

The `MBlock`'s first GUI is used to set block properties such as the numbers, the names and the types of the inputs and outputs, and the parameters. See Figure 14.6. The "I"

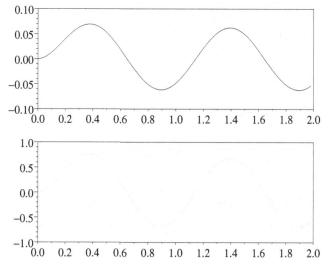

Figure 14.5. Simulation result.

designates an implicit port (input or output), and "E" an explicit port. The blocks parameters contain both the Modelica parameters and variables; the distinction is made by setting the proper parameter properties. The value 0 used here indicate that R and L are both Modelica parameters. Note that unlike the `CBlock4`, `Scifunc` and `GENERIC` blocks, the inputs, outputs, variables and parameters have names associated with them in `MBlock`. The reason is that the Modelica code refers to these variables explicitly.

Based on the information provided by the user in the first GUI box, `MBlock` generates two other GUI boxes. The second GUI box is used to set the numerical values of the parameters defined in the first GUI box. The last GUI is used to write the Modelica code of the block. A portion of the code is generated automatically based on the information provided in the first GUI box.

74 Set Block properties	☐ ▣ ✕
Set Modelica generic block parameters	
Input variables:	"u1"
Input variables types:	"I"
Output variables:	["y1";"y2"]
Output variables types:	["I";"E"]
Parameters in Modelica:	["R";"L"]
Parameters properties:	[0;0]
Function name:	generic
Dismiss	OK

Figure 14.6. Example of an `MBlock`'s first GUI box.

The following example allows us to illustrate how this block can be used to model simple hybrid systems in Modelica using `MBlock`. We use the same bouncing ball model introduced in Chapter 9. The block we want to construct has one explicit output giving the position of the ball. The equations of motion when the ball is in the air are given in

Equations (9.8) and the velocity is reversed when the ball hits the ground. The first GUI box is filled as illustrated in Figure 14.7. As we can see, the block has no input, has only one output, which is explicit, one parameter and two variables with fixed initial values.

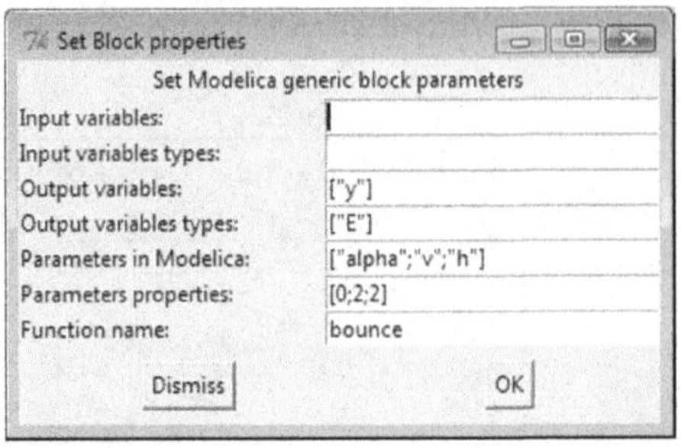

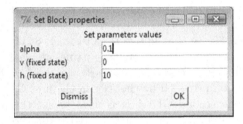

Figure 14.7. MBlock's first GUI box for the bouncing ball example.

The second GUI box is used to set the value of the parameter and the initial values of the states. See Figure 14.8. The code is written in the last GUI box; see Figure 14.9. Note that all the parameter/variable declarations are generated automatically.

Figure 14.8. MBlock's second GUI box for the bouncing ball example.

The **when** keyboard is used in Modelica for generating events. In this case an event is generated at the time instant when **h<0** becomes true. This (zero-crossing) event activates the equations in the body of the **when** statement, which in this case contains only a **reinit** instruction. This **reinit** instruction re-initializes the state v to -v. The code used in this example is standard Modelica code. The latest Modelica specification document can be found at **http://www.modelica.org**. Note however that Scicos does not cover the full extent of the Modelica language specification.

14.6 Initialization

Unlike for systems of ordinary differential equations, the concept of state in differential-algebraic systems is fairly involved as pointed out in Sections 3.1.4 and 3.2.4. In the case of ordinary differential equations, the initialization is usually done by simply furnishing the state x at the initial time. In the case of differential-algebraic systems, both state x and its time derivative $\dot{x}$ must be given. Finding consistent pairs of $(x, \dot{x})$ can be a difficult problem for large systems.

Figure 14.9. MBlock's third and last GUI box for the bouncing ball example.

Scicos provides an interactive tool to help users find consistent initial conditions for Modelica models. The `Modelica Initialize` menu becomes active only when the Scicos model containing Modelica components is compiled. The interactive GUI provided in this tool allows the user to select the variables (states x and their derivatives $\dot{x}$) that are fixed, by giving the weight 1, and the unknown variables, i.e., those that can be modified to obtain a consistent initial condition, specified by weight 0. In the current version of Scicos in ScicosLab 4.4, only weights 0 and 1 are allowed, and the system must be square. This means that when the weight of one variable is set to 1, another must be set to 0.

Note that even for ordinary differential equations, initialization is required if we want to impose constraints on the initial derivatives of the states. One very common application is starting the simulation in steady state.

14.6.1 Example

We consider a simple electrical circuit depicted in Figure 14.10. The simulation result is given in Figure 14.11.

To start the simulation in steady state, after compiling the diagram, the `Modelica initialize` menu is activated. This opens up a first GUI (see Figure 14.12) which allows the user to choose between initialization where all the states are given, all the state derivatives given, or a mixed mode where a set of states and state derivatives is to be specified by the user later. In this case, we select `Fix derivatives`.

This opens up the GUI in Figure 14.13. Having chosen the `Fix derivatives` option, the weight on all state derivatives is set to 1. The values of state derivatives is set by default to zero.

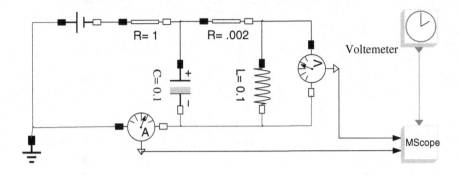

Figure 14.10. Scicos diagram representing a simple electrical circuit.

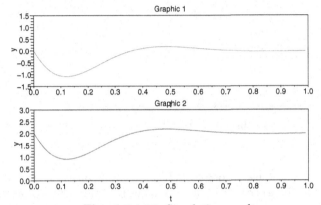

Figure 14.11. Simulation result.

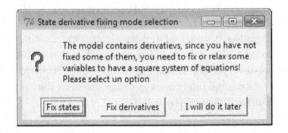

Figure 14.12. The first GUI for initialization of Modelica models.

A number of solvers are proposed by the GUI. After selecting the solver, clicking on the `Solve` button finds the consistent initial condition which in this case, as it can be seen in the GUI, sets the initial current in the inductor (the state derivative) to approximately -2. The simulation following this initialization yields constant signals as expected.

In more complex problems the initial conditions may not be given directly as a a subset of the state variables or their derivatives. Rather they are given in terms of functions of these variables. For example, the total energy could be fixed, or we may have that the position lies on some sort of surface. In some cases these problems can be transformed with some effort into a form that is then solvable by the initialization GUI. We will not

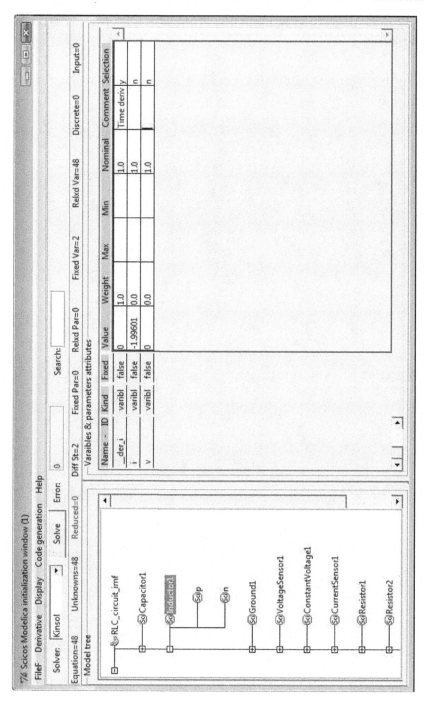

Figure 14.13. The second and main GUI after initialization.

discuss these more complicated problems here. The reader is referred to the DAE literature [11, 12, 15, 35, 55].

A

Inside Scicos

In this chapter, we will give an overview of how the Scicos compiler and editor are implemented. This information is useful for advanced users who want to customize Scicos or develop exotic functionalities. It is also useful for developers of interfaces to other software. But it can also be of interest to the average user, in particular, for efficiently using the debugging facilities.

Our objective here is not to give a detailed description of the underlying code but rather to give enough information so that the code (which of course is available in the source version of Scilab) becomes comprehensible. This requires explaining the data structures used and giving the structure of the main programs.

The data structures used in Scicos are all Scilab variables. Thus they can be examined and altered using standard Scilab functions. There are in particular two variables associated with a Scicos diagram that are of great importance:

- `scs_m` contains all the information concerning the Scicos diagram. It is constructed by the editor.
- `%cpr` contains the result of the compilation; it is used by the simulator.

We start with a brief presentation of the editor and the way `scs_m` is constructed. We then look at the compilation phase, where `scs_m` is used to obtain `%cpr`. `%cpr` is used by the simulator to perform simulation.

A.1 Scicos Editor

The Scicos editor is written entirely in the Scilab language, including the graphical user interface, which uses standard Scilab graphics. This facilitates customization of the editor by the user. This section provides the necessary information.

A.1.1 Main Editor Function

The main Scilab function implementing the Scicos editor is `scicos`, which can be found in file `SCI/macros/auto/scicos.sci`. This function takes as input argument a Scicos diagram (`scs_m`) and *opens* it. Invoked without any argument, it opens an empty diagram. This function is called recursively when a Super block is opened.

The Scicos editor can be parameterized by initializing the following variables:

S.L. Campbell et al., *Modeling and Simulation in Scilab/Scicos with ScicosLab 4.4*,
DOI 10.1007/978-1-4419-5527-2_15, © Springer Science+Business Media, LLC 2010

- `scicos_pal`: list of palettes. An $n \times 2$ matrix of strings, where the first column contains the names of the palettes and the second the file (`.cos` or `.cosf`) in which the palette is defined. In fact, a palette is nothing but a Scicos diagram.
- `%scicos_menu`: Scicos menus. A Scilab list including vectors of strings, where the first element of each vector contains the name of the menu, and the rest, the corresponding items of the menu. Each item corresponds to an operation (such as move, copy, save) and has a corresponding Scilab function.
- `%scicos_short`: keyboard shortcuts. An $n \times 2$ matrix of strings, where the first column contains the character to be used for the shortcut (only lower-case characters can be used) and the second, the name of an operation, which must be an item in one of the menus.
- `%scicos_help`: manual pages associated with operations. A Scilab `tlist` where for each operation, a vector of strings contains the corresponding manual page.
- `modelica_libs`: a vector of strings containing the list of directories (full path), where implicit Modelica blocks are defined.
- `scicos_pal_libs`: a vector of strings containing the list of directories inside `SCI/macros/scicos_blocks` to be loaded. These directories contain the interfacing functions of the blocks in Scicos palettes. The user should not, in general, edit this variable.
- `%scicos_lhb_list`: Scicos contextual popup menus following a mouse right click. The Scilab list contains three elements. `%scicos_lhb_list(1)` contains a vector of strings corresponding to existing menus. This vector is used to construct the list of menu actions available when right clicking on a block. `%scicos_lhb_list(2)` specifies the menu actions available when right clicking on the background with no selected object. Finally, `%scicos_lhb_list(3)` gives the list of menu actions associated with right clicking on a block in a palette.

These variables are initialized by the function `initial_scicos_tables`, which can be found in the directory `SCI/macros/util`. This function is executed as a script by `scilab.star`, which is executed at the startup of Scilab. They can be modified by the user subsequently.

The function `scicos` contains the main event loop of the editor. When an item in one of the menus is invoked to perform an operation, `scicos` calls the corresponding Scilab function. The name of the Scilab function is derived from the name of the menu item by removing special characters "/", ".", and "-", and adding a trailing "_". For example, the function corresponding to the item `Open/set` in the `Object` menu is `OpenSet_`. Functions associated with editor operations have no input-output arguments and are executed as scripts. Scicos comes with default menus providing many elementary editor operations. The associated functions can be found in the directory `SCI/macros/scicos`.

Examples

Here we give examples of adding a new operation to the Scicos editor.

The first one is is just a toy example and consists in analyzing the diagram and displaying the number of blocks, the number of Super blocks, and the number of links in the diagram. We start by adding a new item, which we call `Block Info`. We place it in the last position. For that, the Scilab variable `%scicos_menu` should be modified. This should be done before Scicos is launched or by activating the Scilab window from Scicos.

Default Scicos menus are `File`, `Diagram`, `Palette`... The eighth menu is `Tools`. So if we want to place this item in `Tools`, we should add it to the vector `%scicos_menu(8)`. For example, it can be placed at the end of this menu. Similarly, this item can be added to

the contextual menu. In particular it should be added %scicos_lhb_list(2) because the
action concerns the diagram and not a particular block.

By doing so, the item Diagram Info is added to the Misc menu in Scicos, and when
the user selects this item, Scicos attempts to execute the function DiagramInfo_. So we
need to define this function as well. To implement fully the debugging menu, the following
script can be used.

These instructions must be executed before Scicos is launched. A convenient place for
them is in the .scilab file.

Note that in order to explore the scs_m structure, we examine each item in the objs
field and increment the corresponding counters. When a Super block is encountered, the
function do_diagram_Info is called recursively. This way we can explore the content of
the hierarchical structure scs_m.

The Cmenu variable sets the next operation to be performed. By default in Scicos, the
editor is in Open/Set operation mode; it can also be set to an empty matrix.

The second example consists of adding a Nyquist menu, which displays the nyquist
plot associated with the linearized model of a selected Super block. For this operation,
we use the function lincos introduced in Section 11.3.2. The following is a very simple
implementation that uses the Scilab variables Select, curwin and scs_m used in scicos:

```
[scicos_pal, %scicos_menu, %scicos_short, %scicos_help,0...
        %scicos_display_mode, modelica_libs,scicos_pal_libs,0...
        %scicos_lhb_list, %CmenuTypeOneVector,%scicos_gif,0...

%scicos_menu(8)($+1)="Nyquist";
%scicos_lhb_list(1)($+1)="Nyquist";

 function Nyquist_()
   if size(Select,1)==1 & Select(1,2)==curwin  then
    o=scs_m.objs(Select(1,1));
    if typeof(o)<>"Block"|o.model.sim(1)<>'super'  then
     message('Selected object is not a Super block.')
    else
     sysl=lincos(o.model.rpar);
     h_save=gcf();
     scf(max(winsid())+1);  // open new window
     nyquist(sysl)
     scf(h_save);  // restore editor's window
    end
   else
    message('Select one and only one object in current window.');
   end
Cmenu=[];
 endfunction
```

The variable Select is a two-column matrix holding the object number and the window
number associated with each selected object (and thus highlighted in the editor). The
variable curwin contains the window number of the current editor window (the editor
window having focus). The variable scs_m codes the diagram associated with the current
editor window.

Editor operations are, in general, more complex and involve additional Scilab variables defined and used in `scicos`. In most cases, these operations may modify `scicos` variables such as `scs_m`. The best way to learn about these operations is by examining source codes of the functions associated with default operations.

A.1.2 Structure of `scs_m`

`scs_m` is a Scilab object of type `diagram` having two entries:

- `props`: Diagram properties. Scilab object of type `params`,
- `objs`: list of objects included in the Scicos diagram.

Diagram Properties `params`

The diagram properties are:

- `wpar`: This vector is not currently used. It may be used in the future to code window sizes of the editor.
- `title`: A string containing the name of the diagram. The default value is `"Untitled"`
- `tol`: A vector containing simulation parameters including various tolerances used by the solver:
  - `atol`: Integrator absolute tolerance for the numerical solver.
  - `rtol`: Integrator relative tolerance for the numerical solver.
  - `ttol`: Tolerance on time. If an integration period is less than `ttol`, the numerical solver is not called.
  - `deltat`: Maximum integration time interval. If an integration period is larger than `deltat`, the numerical solver is called more than once in such a way that for each call the integration period remains below `deltat`.
  - `scale`: Real-time scaling; the value 0 corresponds to no real-time scaling. It associates a Scicos simulation time to the real time in seconds. A value of 1 means that each Scicos unit of time corresponds to one second.
  - `solver`: Choice of numerical solver. The value 0 implies `LSODAR` and 100 implies `DASKR`.
  - `hmax`: Maximum step size for the numerical solver. 0 means no limit.
  The default value is `[0.0001,1.000E-06,1.000E-10,100001,0,0]`.
- `tf`: Final integration time. The simulation stops at this time. The default value is 100000.
- `context`: A vector of strings containing Scilab instructions defining variables to be used inside block GUIs. All valid Scilab instructions can be used but not comments.
- `void1`: Not used.
- `options`: Scilab object of type `scsopt` defining graphical properties of the editor such as background color and link color. The fields are the following:
  - `3D`: A list with two entries. The first one is a boolean indicating whether or not blocks should have 3D aspect. The second entry indicates the color in the current colormap to be used to create the 3D effect. The default is 33, which corresponds to gray added by Scicos to the standard colormap, which contains 32 colors. The default value is `list(%t,33)`.
  - `Background`: Vector with two entries: background and foreground colors. The default value is `[8,1]`.

– Link: Default link colors for regular and activation links. These colors are used only at link construction. Changing them does not affect already constructed links. The default value is [1,5], which corresponds to black and red if the standard Scilab colormap is used.
– ID: A list of two vectors including font number and sizes. The default value is list([5,1],[4,1]).
– Cmap: An $n \times 3$ matrix containing RGB values of colors to be added to the colormap. The default value is [0.8,0.8,0.8], i.e., the color gray.

- void2: Not used .
- void3: Not used.
- doc: Used for documenting the diagram.

Diagram Content

The field objs contains a Scilab list of the objects within the diagram. The objects can be of type Block, Link, or Text. A Block can be a basic block or a Super block.

Scicos Block

It is a structure including the following fields.

- graphics: Scilab object of type graphics including graphical information concerning the features of the block. The fields are:
  – orig: Vector [xo,yo], where xo is the x coordinate of the block origin and yo is the y coordinate of the block origin.
  – sz: Vector [w,h], where w is the block width and h the block height.
  – flip: Boolean indicating block orientation. It is used to switch the ports on the lef- hand side and those on the right-hand side of the block.
  – exprs: A vector of strings including formal expressions (usually including numbers and variable names) used in the dialog of the block.
  – pin: Vector with pin(i), the number of the link connected to the ith regular input port (counting from one), or 0 if this port is not connected.
  – pout: Vector with pout(i), the number of the link connected to the ith regular output port (counting from one), or 0 if this port is not connected.
  – pein: Vector with pein(i), the number of the link connected to the ith event input port (counting from one), or 0 if this port is not connected.
  – peout: Vector with peout(i), the number of the link connected to the ith event output port (counting from one), or 0 if this port is not connected.
  – gr_i: A list containing two elements. The first element is a vector of strings including Scilab graphics expressions for drawing block's icon, and second, an integer indicating the background color of the block.
  – id: A string including an identification for the block. The string is displayed underneath the block in the diagram.
  – in_implicit: A vector of strings including "E" and "I". E and I stand respectively for explicit and implicit port, and this vector indicates the nature of each input port. For regular blocks (not implicit), this vector is empty or contains only E's.
  – out_implicit: Similar to in_implicit but for the output ports.
- model: Scilab object of type model including the following fields.

- **sim:** A list containing two elements. The first element is a string containing the name of the computational function (C, Fortran, or Scilab). The second element is an integer specifying the type of the computational function. Currently type 4 and 5 are used, but older types continue to work to ensure backward compatibility. The reserved function names **super** and **csuper** are used to indicate that the block is respectively a Super block or a compiled Super block.
- **in:** A vector specifying row sizes of regular inputs.
- **in2:** A vector specifying column sizes of regular inputs.
- **intyp:** A vector specifying the types of regular inputs.
- **out:** A vector specifying row sizes of regular outputs.
- **out2:** A vector specifying column sizes of regular outputs.
- **outtyp:** A vector specifying the types of regular outputs.
- **evtin:** A vector specifying the number and sizes of activation inputs. Currently activation ports can be only of size one.
- **evtout:** A vector specifying the number and sizes of activation outputs.
- **state:** Vector containing initial continuous-time state.
- **dstate:** Vector containing initial discrete-time state.
- **odstate:** List of matrices having different types and sizes.
- **rpar:** In case of a basic block, it is a vector of real parameters passed to associated computational function. In the case of a Super block (compiled or not), it contains the **scs_m** structure of the diagram enclosed.
- **ipar:** Vector of integer parameters passed to associated computational function.
- **opar:** List of matrices having different types and sizes.
- **blocktype:** It can be set to **c** or **d** indifferently for regular blocks. **x** is used if we want to force the computational function to be called during the simulation phase even if the block does not contribute to computation of the state derivative.
- **firing:** Vector of initial event firing times of size equal to the number of activation output ports. A value ≥ 0 programs an activation (event) at the corresponding port to be fired at the specified time.
- **dep_ut:** Boolean vector [**timedep udep**].
    - **timedep** boolean: true if block is *always active*.
    - **udep** boolean vector: ith element is true if the block has a direct feed-through from the ith input to one of the outputs, i.e., at least one of the outputs depends directly (not through the states) on the ith input. If all inputs are identical, a scalar can be used.
- **label:** A string. The label can be used to identify a block in order to access or modify its parameters during simulation.
- **nzcross:** Number of zero-crossing surfaces.
- **nmode:** Number of modes. Note that this gives the size of the vector mode and not the total number of modes in which a block can operate in. Suppose a block has 3 modes and each mode can take two values, then the block can have up to $2^3 = 8$ modes.
- **equations:** Used in case of Implicit blocks; see Chapter 14.
- **gui:** The name of the Scilab GUI function associated with the block.
- **doc:** Used for documentation of the block.

The coding used for the data types is the following:

1   Scilab real matrix
2   Scilab complex matrix
3   Scilab int32 matrix

4 Scilab int16 matrix
5 Scilab int8 matrix
6 Scilab uint32 matrix
7 Scilab uint16 matrix
8 Scilab uint8 matrix
9 Scilab Boolean matrix
-1 Any other Scilab variable

These data types can be used in **odstate**, **opar**, and, except for the type -1, for inputs and outputs. The type -1 is mainly used for blocks having computational functions written in Scilab.

Scicos Link.

It is a Scilab list including the following fields:

- **xx:** A vector. A link is defined as a polyline line. **xx** defines the x-coordinate of the points characterizing the polyline.
- **yy:** A vector having the same size as **xx**. It defines the y-coordinate of the points characterizing the polyline.
- **id:** A string corresponding to the name of the function drawing the link. Default value is "drawlink".
- **thick:** Vector of size two defining line thickness.
- **ct:** A vector. The first entry designates the color, and the second, the nature. The second entry is 1 for a regular link, -1 for an activation link, and 2 for an implicit link.
- **from:** Vector of size three including the block number, port number, and port type (0 for output, 1 for input) at the origin of the link. Note that the third entry may be 1 if the link is implicit; otherwise it is zero.
- **to:** Vector of size three including the block number, port number, and port type at the destination of the link.

Once the diagram has been successfully edited, **scs_m** can be passed on to the compiler.

A.2 Scicos Complier

Scicos diagram compilation is done in two stages (there is an additional precompilation phase if the diagram contains implicit blocks; see Chapter 14). These two stages are implemented by the Scilab functions **c_pass1** and **c_pass2**.

A.2.1 First Compilation Stage

The first stage of the compilation consists in removing the hierarchy from the diagram and constructing a flat description. This is done by **c_pass1**, which has the following calling sequence:

```
[blklst,cmat,ccmat,cor,corinv,ok]=c_pass1(scs_m)
```

- **blklst** is a list of blocks present in the diagram. It contains block information relevant to simulation. Block properties such as color, icon, size, and location, which are not useful for simulation, have been stripped.

- cmat is an $n \times 6$ matrix. Each row corresponds to a regular link and contains the block number, port number, and port type (explicit or implicit) of the source block and the same information regarding the destination block.
- ccmat is an $n \times 4$ matrix. Each row corresponds to an activation link and contains the block number and port number of the source block and the block number and port number of the destination block.
- cor and corinv are correspondence tables (coded as lists) used to find the correspondence between blocks in blklst and scs_m.

A.2.2 Second Compilation Stage

The second stage is done by the main compilation function c_pass2, which constructs all the scheduling tables and other information needed for simulation and code generation. The calling sequence is as follows:

```
%cpr=c_pass2(blklst,connectmat,ccmat,cor,corinv)
```

The input arguments are all generated by c_pass1. connectmat is simply cmat, but the third and last columns have been removed. The output %cpr is a Scilab structure containing all the information needed by the simulator. We will not explain how c_pass2 computes %cpr. Instead we give a detailed description of %cpr in the next section.

A.2.3 Structure of %cpr

The Scilab object %cpr contains the result of the compilation. The simulator uses only %cpr. It is thus important for an advanced user to understand how compilation results are coded in %cpr.

- state: Scilab object of type xcs. It contains all the states of the model, that is, everything than can evolve during the simulation. It contains in particular
  - x: The continuous-time state, which is obtained by concatenating the continuous-time states of all the blocks.
  - z: The discrete-time state, which is obtained by concatenating the discrete-time states of all the blocks.
  - iz: Vector of size equal to the number of blocks. If a block needs to allocate memory at initialization, the associated pointer is saved here.
  - tevts: Vector of size equal to the number of activation sources. It contains the scheduled times for programmed activations.
  - evtspt: Vector of size equal to the number of activation sources. It is an event scheduler.
  - pointi: The number of the next programmed event.
  - outtb: Vector containing all the link memories. Link memories hold block output values.
- sim: Scilab object of type scs. It contains information that does not evolve during the simulation.
  - funs: A vector containing the name of the computational functions.
  - xptr: A vector pointer to the continuous time state x. The continuous-time state of block i is
    %cpr.state.x(%cpr.sim.xptr(i):%cpr.sim.xptr(i+1)-1).
  - zptr: A vector pointer to the discrete-time state z, which is similar to xptr.
  - zcptr: A vector pointer to the zero-crossing surfaces.

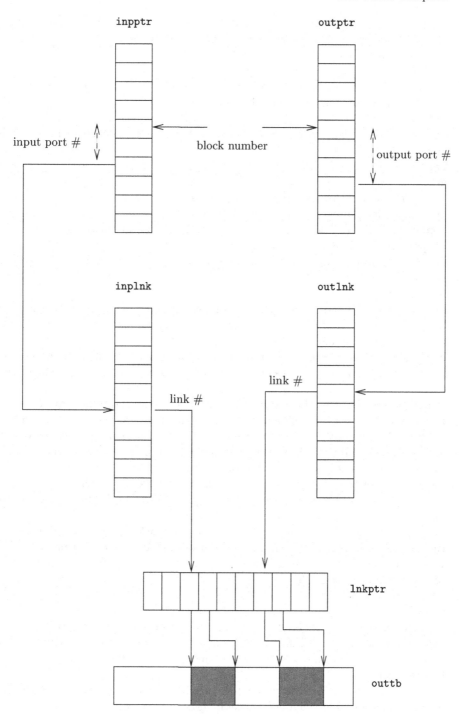

Figure A.1. `outtb` contains the link registers. The memory allocated to link `l` is `outtb([lnkptr(l):lnkptr(l+1)-1])`. The link number of the link connected to input `i` of block `j` is `l=inplnk(inpptr(j)+i-1)`. Similarly, the link number of the link connected to output `i` of block `j` is `l=outlnk(outptr(j)+i-1)`.

- inpptr: A vector pointer used to find the link number and consequently the part of outtb corresponding to a given input port. See Figure A.1.
- outptr: Similar to inpptr but for output ports.
- inplnk: Similar to inpptr. See Figure A.1.
- outlnk: Similar to inplnk but for output ports.
- lnkptr: A vector pointer to outtb indicating the region corresponding to a given link.
- rpar: Vector of real parameters that is obtained by concatenating the real parameters of all the blocks.
- rpptr: A vector pointer to real parameters rpar. The real parameters of block i are
  %cpr.sim.rpar(%cpr.sim.rpptr(i):%cpr.sim.rpptr(i+1)-1).
- ipar: Vector of integer parameters, similar to rpar.
- ipptr: A vector pointer to integer parameters, similar to rpptr.
- clkptr: A vector pointer to output activation ports.
- ordptr: A vector pointer to ordclk designating the part of ordclk corresponding to a given activation.
- execlk: Not used.
- ordclk: An $n \times 2$ matrix associated to blocks activated by output activation ports. The first column contains the block number, and the second, the event code by which the block should be called. See Figure A.2.
- cord: An $n \times 2$ matrix associated to always active blocks. The first column contains the block number, and the second, the event code by which the block should be called.
- oord: Subset of cord, which affects the continuous-time state derivative.
- zord: Subset of cord, which affects the computation of zero-crossing surfaces.
- critev: A vector of size equal to the number of activations and containing zeros and ones. The value one indicates that the activation is critical in the sense that the continuous-time solver must be cold restarted.
- nb: Number of blocks. Note that the number of blocks may differ from the original number of blocks in the diagram because c_pass2 may duplicate some conditional blocks.
- ztyp: A vector of size equal to the number of blocks. A 1 entry indicates that the block may have zero-crossings, even if it doesn't in the context of the diagram. Usually not used by the simulator.
- nblk: Not used. Set to nb.
- ndcblk: Not used.
- subscr: Not used.
- funtyp: A vector of size equal to the number of blocks indicating the type of the computational function of the block. Block type can be 0 through 5. Currently only type 4 (C language) and type 5 (Scilab language) computational functions should be used. But older blocks can also be used.
- iord: An $n \times 2$ matrix associated to blocks that must be activated at the start of the simulation. This includes blocks inheriting from constant blocks and always active blocks.
- labels: A string vector of size equal to the number of blocks containing block labels.
- modptr: A vector pointer to the block modes.

- cor: Scilab list with a hierarchy identical to that of the original diagram and including block numbers in the compiled structure. It allows finding block number in %cpr from position in scs_m.
- corinv: Scilab list containing vectors. It allows finding position in scs_m from block number in %cpr.

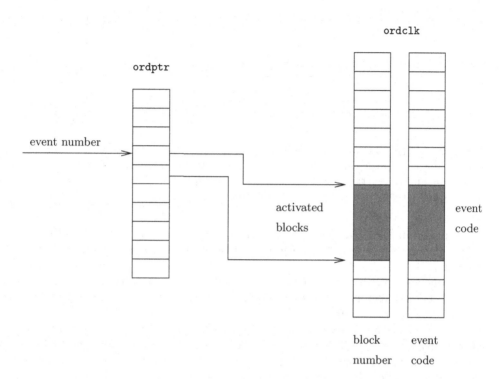

Figure A.2. The table ordptr is a pointer to ordclk, which contains the blocks that are activated by any given activation. For activation i, the associated blocks are found in ordclk([ordptr(i):ordptr(i+1)-1],1).

A.2.4 Partial Compilation

For large diagrams including many conditional blocks, the compilation time can be noticeable. That is why Scicos does not do a complete recompilation if minor modifications are made in a compiled diagram. In most cases (but not always), changing a block parameter does not affect scheduling tables in the compiled structure. In such a case, only a partial compilation is performed, which means that some elements of %cpr such as %cpr.sim.rpar are updated without recomputing scheduling tables %cpr.sim.ordclk, which can be time-consuming.

Anytime a block parameter is modified, Scicos determines the effect of the modification on the compiled structure by setting the level of compilation required. This level is coded in the Scilab variable needcompile. If a diagram is not compiled, needcompile is 4. For a freshly compiled diagram, needcompile is zero. In many cases, changing a block parameter, for example the gain value of the Gain block, does not affect needcompile. The block is

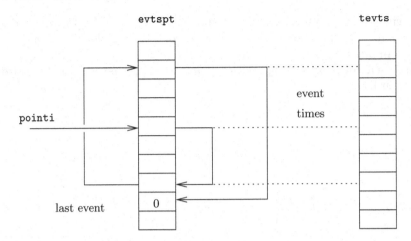

Figure A.3. The table **evtspt** is used to store the programmed events. The variable **pointi** contains the next programmed event number. **i=evtspt(pointi)** is the number of the event programmed next; **j=evtspt(i)** is next and so on. **evtspt** of the last programmed event is zero. Event times for each event are stored in table **tevts**.

simply placed in the variable **newparameters** so that the new parameters of the block can be copied from **scs_m** into **%cpr** before simulation. This is done by the Scilab function **modipar**.

If changing a parameter changes link sizes too, then **needcompile** goes up to one. In this case, pointer vectors may need to be recomputed. However, scheduling tables remain valid and are not recomputed.

If the modification is more complex and involves, for example, preprogrammed activation signals, then it still may not be necessary to recompute scheduling tables. In this case, **needcompile** becomes 2 and Scilab function **c_pass3** recomputes all the entries from **scs_m**, except for scheduling tables, before simulation.

Finally, when it is absolutely necessary, for example following any operation modifying the diagram (adding or deleting a block for example), **needcompile** becomes 4 and a recompilation is performed before simulation.

The Scilab function **do_update** handles the partial compilation. Note that a recompilation can always be imposed by the user by choosing the menu item **Compile**.

A.3 Scicos Simulator

The interface to the Scicos simulator in Scilab is the function **scicosim**. We have already encountered this function in Section 11.1.2. This function is used to initialize, run, and end Scicos simulations. These operations are done by the function **do_run** (in the **SCI/macros/scicos** directory), which is called when the user runs a simulation in Scicos.

The Scicos simulator is a complex C program. We do not study it here. The simulation functions are in the file **SCI/routines/scicos/scicos.c**. The main function is **scicos**, which calls the routines **cosini**, **cossim**, and **cosend** depending on whether initialization, simulation, or ending has been requested.

B

Coding Examples

This appendix gives several examples of coding using different features of ScicosLab.

B.1 Scicos Blocks of Type 5

We present here two examples of Scicos blocks using type 5 computational functions.

B.1.1 Type 5 Block for the Bouncing Ball Example

This section gives the interfacing and the computational functions of a type 5 block realizing the bouncing ball example presented on page 212. In this case, the full dynamics is included in a basic block.

Interfacing Function

The interfacing function for the block realizing the dynamics of the bouncing ball problem is given below. The user can change the initial position and velocity of the ball.

```
function [x,y,typ]=BouncingBall5(job,arg1,arg2)
 x=[];y=[];typ=[];
  select job
   case 'plot' then
   standard_draw(arg1)
   case 'getinputs' then
   [x,y,typ]=standard_inputs(arg1)
   case 'getoutputs' then
   [x,y,typ]=standard_outputs(arg1)
   case 'getorigin' then
   [x,y]=standard_origin(arg1)
   case 'set' then
   x=arg1
   graphics=arg1.graphics;
   exprs=graphics.exprs;
   model=arg1.model;
    while %t do
      [ok,xx,rpar,exprs]=..
```

```
            getvalue('Set BOUNCEBALL block parameters',..
                ['Initial ball position and speed';..
                 'Parameter alpha'],..
                list('vec',2,'vec',1),exprs)
      if ok  then
        model.rpar=rpar;model.state=xx
        graphics.exprs=exprs
        x.graphics=graphics;x.model=model
      end
      break
    end
  case 'define'  then
  model=scicos_model()
  model.sim=list('bounceball5',5);
  model.out=1
  model.state=[2;0]
  model.rpar=1
  model.nzcross=1;
  model.blocktype='c';
  model.dep_ut=[%t %t];
  exprs=[sci2exp(model.state);sci2exp(model.rpar)]
  gr_i=['xstringb(orig(1),orig(2),''BOUNCEBALL'',sz(1),sz(2),''fill'');']
  x=standard_define([3 2],model,exprs,gr_i)
  end
endfunction
```

Note that the first entry of dep_ut, set to true, is not used because the block has no input, but the second entry specifies that the block is always active.

Computational Function

This computational function should be compared to the C function Bounceball defined on page 213.

```
function block=bounceball5(block,flag)
  if flag==1
    block.outptr(1)(1)=block.x(1)
  elseif flag==0
    block.xd(1)=block.x(2);
    block.xd(2)=-9.8-block.rpar(1)*block.x(2)^3
  elseif flag==2&block.nevprt==-1&block.jroot(1)<0
    block.x(2)=-block.x(2)
  elseif flag==9
    block.g(1)=block.x(1)
  end
endfunction
```

B.1.2 Animation Block for the Cart Pendulum Example

Animation of simulation results is often helpful both in understanding the results and in presenting them to others. In this section we give the interfacing and the computational functions associated with the animation block of the cart pendulum problem presented in

Section 11.3. The computational function of this block is type 5; that is, a Scilab program is used to perform the animation during the simulation.

This block receives the position of the cart z and the angle of the pendulum θ as input. The angle ϕ, which is a constant, is a block parameter as are the sizes of the cart and the pendulum and the coordinate of the viewing window. These parameters are initialized and adjusted by the interfacing function below:

```
function [x,y,typ]=PENDULUM_ANIM(job,arg1,arg2)
// Animation of the cart-pendulum problem
  x=[];y=[];typ=[]
  select job
    case 'plot' then
    standard_draw(arg1)
    case 'getinputs' then
    [x,y,typ]=standard_inputs(o)
    case 'getoutputs' then
    x=[];y=[];typ=[];
    case 'getorigin' then
    [x,y]=standard_origin(arg1)
    case 'set' then
    x=arg1;
    graphics=arg1.graphics;exprs=graphics.exprs
    model=arg1.model;dstate=model.dstate
    while %t do
      [ok,plen,csiz,phi,xmin,xmax,ymin,ymax,exprs]=getvalue(..
        'Set Scope parameters',..
        ['pendulum length';'cart size (square side)';'slope';
        'Xmin';'Xmax'; 'Ymin'; 'Ymax'; ],..
        list('vec',1,'vec',1,'vec',1,'vec',1,'vec',1,'vec',1,'vec',1),exprs)
      if ~ok then break, end
    mess=[]
      if plen<=0|csiz<=0 then                        .
        mess=[mess;'Pendulum lenght and cart size must be positive.';' ']
        ok=%f
      end
      if ymin>=ymax then
        mess=[mess;'Ymax must be greater than Ymin';' ']
        ok=%f
      end
      if xmin>=xmax then
        mess=[mess;'Xmax must be greater than Xmin';' ']
        ok=%f
      end
      if ~ok then
        message(mess)
      else
        rpar=[plen;csiz;phi;xmin;xmax;ymin;ymax]
        model.rpar=rpar;
        graphics.exprs=exprs;
        x.graphics=graphics;x.model=model
         break
      end
    end
  end
```

```
      case 'define'  then
      plen=2; csiz=2; phi=0;
      xmin=-5;xmax=5;ymin=-5;ymax=5
      model=scicos_model()
      model.sim=list('anim_pen',5)
      model.in=[1;1]
      model.evtin=1
      model.dstate=0
      model.rpar=[plen;csiz;phi;xmin;xmax;ymin;ymax]
      model.blocktype='d'
      model.dep_ut=[%f %f]
      exprs=string(model.rpar)
      gr_i=['thick=xget(''thickness'');xset(''thickness'',2);';
             'xx=orig(1)+sz(1)*[.40.60.60.40.4]'
             'yy=orig(2)+sz(2)*[.20.20.40.40.2]'
             'xpoly(xx,yy,''lines'')'
             'xx=orig(1)+sz(1)*[.50.6]'
             'yy=orig(2)+sz(2)*[.40.8]'
             'xpoly(xx,yy)'
             'xset(''thickness'',thick);']
    x=standard_define([3 3],model,exprs,gr_i)
  end
endfunction
```

Note that this block has 2 regular scalar inputs and an activation input. The activation input receives during the simulation the information concerning the time instants when the display has to be updated. Note also that the block has a discrete-time state, set to zero. The reason for the presence of this state is to make sure that during the simulation, the block is called with `flag` equal to 2. Since this block has no output (from the point of view of Scicos; there is indeed an output to the screen), the block is never called with `flag` equal to 1. So in the absence of a state, the block is never called during simulation. Adding a state is one way to overcome this problem. The other is to add outputs to the block. This latter method works even if these outputs are left unconnected. Here we have chosen to add a state, which turns out to be useful for other reasons.

The computational function of this block is defined in PENDULUM_ANIM to be `anim_pen`. This function is given below.

```
function [blocks] = anim_pen(blocks,flag)
  win=20000+curblock()
  if flag<>4  then H=scf(win),  end
  xold=blocks.z
  rpar=blocks.rpar
  plen=rpar(1);csiz=rpar(2);phi=rpar(3);
  if flag==4  then
    xset("window",win)
    set("figure_style","new")
    H=scf(win)
    clf(H)
    H.pixmap='on'
    Axe=H.children
    Axe.data_bounds=rpar(4:7)
    Axe.isoview='on'
```

```
    S=[cos(phi),-sin(phi);sin(phi),cos(phi)]
    XY=S*[rpar(4),rpar(5);-csiz/2,-csiz/2]
    xsegs(XY(1,:),XY(2,:))
    x=0;theta=0;
    x1=x-csiz/2;x2=x+csiz/2;y1=-csiz/2;y2=csiz/2
    XY=S*[x1 x2 x2 x1 x1;y1,y1,y2,y2,y1]
    xpoly(XY(1,:),XY(2,:),"lines",1)
    XY=S*[x,x+plen*sin(theta);0,0+plen*cos(theta)]
    xsegs(XY(1,:),XY(2,:))
  elseif flag==2 then
    Axe=H.children
    x=blocks.inptr(1)(1)
    theta=blocks.inptr(2)(1)
    XY=Axe.children(2).data'+..
       [cos(phi)*(x-xold);sin(phi)*(x-xold)]*ones(1,5)
    Axe.children(2).data=XY'
    x1=x*cos(phi);y1=x*sin(phi)
    XY=[x1,x1+plen*sin(theta);y1,y1+plen*cos(theta)]
    Axe.children(1).data=XY'
    blocks.z=x
    show_pixmap()
  end
endfunction
```

This block uses Scilab's new graphics (in contrast to Scicos, which still uses the old graphics). The window number of the graphical display is set to 20000 plus the block number. This is to make sure that if two such blocks are used in the same diagram, they use different windows for their displays. The discrete-time state `blocks.z`, which has been introduced simply to make sure the block is called during simulation, is also used to store the previous position of the cart. This information is then used to update the associated graphics data structure in a slightly more efficient way. This is done by computing the change in the position by subtracting the old position from the new position.

The `isoview` property is used to make sure the horizontal and vertical scales remain proportional on the screen after resizing the window. The `pixmap` mode is used to create an efficient and smooth animation. In particular, the display is refreshed only after all the changes have been done. Note that the block does not ensure real-time simulation. This must be done by adjusting the corresponding simulation parameter of the diagram.

B.2 Animation Program for the Car Example

This section provides the Scilab program used to construct the animation of the car example in Section 5.2. This program can be used as an example for developing animation programs in Scilab using the new graphics mode. Note also the use of the Scilab functions realtime and realtimeinit used to make the animation run in real time.

```
function play(T,X,XI)
  scf(1);clf();h=gcf();
  h.pixmap = "on";
  h.figure_size = [1400 600];
  h.children.axes_visible=['off','off','off'] ;
  h.children.isoview='on';
  h.children.tight_limits = "on"
  h.children.margins = [0,0,0,0];
  XXX=[-XI-12:0.2:3*XI+12];
  FY=[];XC=[];
  for x=XXX,
    th=atan(Fx(x));
    XC=[XC,x+sin(th)*r2]
    FY=[FY,F(x)-cos(th)*r2];
  end
  h.children.data_bounds=[min(XC),min(FY);max(XC),max(FY)+d];
  xfpoly([XC(1),XC,XC($)],[min(FY)-d,FY,min(FY)-d]); hfp = gce();
  hfp.foreground = 5;
  drawlater()
  h.pixel_drawing_mode='xor'
  l3=1.15*l2; // car body length (half) for animation
  realtimeinit(1); // intialize real time operation
  for i=1:size(T,'*')
    t=T(i);
    X=XX(:,i);
    x=X(1);xd=X(2);y=X(3);yd=X(4);th=X(5);thd=X(6);xsi=X(7);eta=X(8);
    x0=x+xsi*sin(th)-l2*cos(th);
    x1=x+eta*sin(th)+l2*cos(th);
    y0=y-xsi*cos(th)-l2*sin(th);
    y1=y-eta*cos(th)+l2*sin(th);
    xp=[x0,x-l2*cos(th);x+l2*cos(th),x1]';
    yp=[y0,y-l2*sin(th);y+l2*sin(th),y1]';
    xpp=[x-l3*cos(th);x+l3*cos(th)];
    ypp=[y-l3*sin(th);y+l3*sin(th)];
    W1=[x0-r2,y0+r2,2*r2,2*r2,0,360*64];
    W2=[x1-r2,y1+r2,2*r2,2*r2,0,360*64];
    if i==1 then
      xpoly(xpp,ypp); hp1 = gce();
      hp1.thickness=10;
      xpolys(xp,yp); hp2 = gce();hp21=hp2.children(1);hp22=hp2.children(2)
      xfarcs([W1',W2']); ha = gce();ha1=ha.children(1);ha2=ha.children(2)
      realtime(t);
      draw([hp1,hp2,ha])//draw car
    else
      draw([hp1,hp2,ha])//erase car
      hp1.data=[xpp,ypp];
```

```
      hp21.data=[xp(:,1),yp(:,1)];
      hp22.data=[xp(:,2),yp(:,2)];
      ha1.data=W1;
      ha2.data=W2;
      realtime(t);
      draw([hp1,hp2,ha])//draw car
    end
    show_pixmap()
  end
 h.pixel_drawing_mode='copy'
 drawnow()
 show_pixmap()
endfunction
```

B.3 Extraction Program for the LaTeX Graphic Example

This function exports to a file in mixed LaTeX and Postscript code the content of a Scilab graphic window. This file has been in particular used in Section 2.4.4 to produce the two files `figps-ps.eps` and `figps-tex.tex`.

```
function xs2latex(win,fname)
  base=basename(fname);
  I=strindex(base,'.');
   if I==[]  then
    fname_eps= fname+'.eps';
    fname= fname+'.ps';
   else
    suf=part(fname,I($):length(fname));
     if suf<>".ps"  then
       error('xs2latex: fname must be ended by0.ps');
       return;
     end
    fname_eps= part(fname,1:I($-1))+'.eps';
   end;
  xs2ps(win,fname)
  unix(SCI+'/bin/Blatexpr 1 1 '+fname);
  txt=mgetl(fname_eps);
  I=grep(txt,"%Latex:");
  txt1=txt(I);
  I=grep(txt1,"nan");
  txt1(I)=[];
  txt1=strsubst(txt1,"%Latex:","");
  mputl(txt1,base+'-tex.tex');
  txt1=txt;
  I=[];
   for i=1:size(txt,'*')
    i,
    l=length(txt(i));
     if l-4 >= 1  then
       if part(txt(i),(l-4):l) == " Show"  then I=[I,i]; end
     end
```

```
       end
     txt1(I)=[];
     mputl(txt1,base+'-ps.eps');
   endfunction
```

B.4 Maple Code Used for Modeling the *N*-Link Pendulum

Here we give the Maple code for the *N*-link pendulum presented in Section 5.1. We have removed from the Maple code functions devoted to LATEX code generation and to energy computation (`ener.f`).

```
read('Euler.map');

#----------------------------------------------------------------------------
# Lagrangian for the N-link pendulum ( l[i]: length of links  r[i]:=l[i]/2;)
#----------------------------------------------------------------------------

# number of links in the pendulum
n:=2;

# [x,y,theta] is of size three
mm:=3:

LLi:=proc(i)
     (1/2)*m[i]*( ( xd[i-1]- r[i] *sin(th[i])*thd[i])**2 +
     ( yd[i-1]+ r[i]*cos(th[i])*thd[i])**2 ) + 1/2*J[i]*(thd[i])**2
     -m[i]*g*(y[i-1]+r[i]*sin(th[i])):
end:

# The point zero is fixed

x[0]:=0:xd[0]:=0: y[0]:=0:yd[0]:=0:

L:=sum( LLi(j),j=1..n):

# Lagrangian variables:
q := [ seq (op([x[i],y[i],th[i]]),i=1..n)]:
qd := [ seq (op([xd[i],yd[i],thd[i]]),i=1..n)]:
qdd:= [ seq (op([xdd[i],ydd[i],thdd[i]]),i=1..n)]:

# Lhs of Euler equations
EL:=euler_equations(L,q,qd,qdd): EE:=map((i)->rhs(i),EL):

#--------------------------------------------------------
# Rewriting the Euler equations to have a canonical form
# used only for output
#--------------------------------------------------------

XX:=CEuler(EE,q,qd,qdd):
ME1:=XX[1]:
ME1:=subs(seq(m[k]*r[k]*cos(th[k])=mrc[k],k=1..n),
     seq(m[k]*r[k]*sin(th[k])=mrs[k],k=1..n),eval(ME1)):
```

```
#--------------------------------------------------
# Constraints on the N-link Pendulum
#--------------------------------------------------

ncont:=2*n;
cont:=[ seq(op([ x[i]-x[i-1] - 2*r[i]*cos(th[i]),
         y[i]-y[i-1] - 2*r[i]*sin(th[i])]),i=1..n)]:

# time derivative of constraints;
cont1:=map ( (exp)->( time_diff(exp,q,qd,qdd)),cont):

# derivatives of constraints are of type Aprim qd = 0
Aprim:=genmatrix(cont1,qd):

#--------------------------------------------------
# Computing S(q);
#--------------------------------------------------

SS:=linsolve(Aprim,matrix(ncont,1,0)):

#------------------------------------------------------------------
# Since the indices can be mixed we have to change SS
# so that the correspondence between the eta[i] and the thd[i]
# will be the identity.  This is convenient, but not essential,
# in order to have a simple interpretation of eta
#------------------------------------------------------------------

permut:=seq(SS[mm*i,1]=t_s[i],i=1..n): SS:=subs(permut,eval(SS)):
permut:=seq(t_s[i]=eta[i],i=1..n):     SS:=subs(permut,eval(SS)):
S:= genmatrix(convert(convert(SS,vector),list),[seq( eta[i],i=1..n)]):

#------------------------------------------------------------
# Left multiplication in the Euler equations.
#------------------------------------------------------------

E1:=multiply(transpose(S),EE):

#------------------------------------------------------------
#  .
# q= S(q) eta ; here  eta = [eta1,eta2,...]
#                          ..
# we use this equation to compute q
#------------------------------------------------------------

qt  := [ seq (eta[i]   ,i=1..n)]: qtd := [ seq (etad[i] ,i=1..n)]:
qtdd:= [ seq (etadd[i],i=1..n)]:
qqdd:=map((x,y,z,t)-> time_diff(x,y,z,t),eval(SS),
     [op(q),op(qt)],[op(qd),op(qtd)],[op(qdd),op(qtdd)]):

#------------------------------------------------------------
#          ..                          .
# Using  q= d/dt [ S(q) eta] and q= S(q) eta
# we can substitute these expressions in E1
```

```
#------------------------------------------------------

E2:=subs(seq(qdd[i]=qqdd[i,1],i=1..nops(qdd)),eval(E1)):
E3:=subs(seq(qd[i]=SS[i,1],i=1..nops(qd)),eval(E2)):


#------------------------------------------------------
# The global system is now
#              .
#  E3 = 0 and q= S(q) eta
#------------------------------------------------------

E3:=map((x)-> simplify(x),E3):


#------------------------------------------------------------------
# Canonical representation for the simplified Euler equations
#              .
# E1= ME(q)  eta + RE(q,eta).
# We use CEuler with a little trick in the parameter call qt,qt,qtd
#------------------------------------------------------------------

XX1:=CEuler(E3,qt,qt,qtd):
MM3:=map((i)->factor(combine(i,trig)),XX1[1]):
RR3:=map((i)->factor(combine(i,trig)),XX1[2]):


#------------------------------------------------------
# FORTRAN GENERATION
# First routine npend(neq,t,th,ydot)
#------------------------------------------------------

flist:=[subroutinem,`npend`,[`neq`,`t`,`th`,`ydot`],
            [
             [ parameterf,[`n=`.n]],
             [ declaref,`implicit doubleprecision`,[`(t)`] ],
             [ declaref,doubleprecision,[`t,th(2*n),eta(n),ydot(2*n),r(n)`]],
             [ declaref,doubleprecision,[`me3s(n,n),const(n,1),j(n),m(n)`]],
             [ declaref,doubleprecision,[`w(3*n),rcond        `]],
             [ declaref,integer,[`i,k,neq,ierr`]],
             [ declaref,`data g`,[`/ 9.81/`]],
             [ declaref,`data r`,[`/ n*1.0/`] ],
             [ declaref,`data m`,[`/ n*1.0/`] ],
             [ declaref,`data j`,[`/ n*0.3/`] ],
             [ dom , `i ` ,1,`n `,1,[ equalf,`ydot(i)`,`th(i+n)`]],
             [ dom , `i ` ,1,`n `,1,[ equalf,`eta(i)`,`th(i+n)`]],
             [ matrixm,`me3s`,MM3 ] ,
             [ matrixm,`const`, RR3 ] ,
             [ dom , `i ` ,1,`n `,1,[ equalf,`const(i,1)`,`-const(i,1)`]],
             [commentf,` solving M z = const to obtain ydot((n+1)..2*n)`],
             [ callf , `dlslv`,[`me3s,n,n,const,n,1,w, rcond,ierr,1`]],
             [ if_then_m,ierr<>0,[
               [writef,6,ff_w,[]],
               [formatf,ff_w,[`'Matrix is badly conditioned'`]]]],
             [ dom , `i ` ,1,`n `,1,[ equalf,`ydot(n+i)`,`const(i,1)`]],
             [returnf]]]:
```

```
Gener('npend.f',flist):

# New np.f
flist:=[subroutinem,'np',['i '],
        [[ declaref,integer,['i ']],[ equalf,'i ',n],[returnf]]]:

Gener('np.f',flist):
```

References

1. Y. Alain. *Théorie et analyse du signal. Cours et initiation pratique via Matlab et Scilab.* Ellipses, Paris, 1999.
2. G. Allaire and S. Kaber. *Algèbre linéaire numérique. Cours et exercices.* Ellipses, Paris, 2002.
3. G. Allaire and S. Kaber. *Introduction á Scilab - Exercices pratiques corrigés d'algèbre linéaire.* Ellipses, Paris, 2002.
4. E. Anderson, Z. Bai, C. Bischof, S. Blackford, J. Demmel, J. Dongarra, J. D. Croz, A. Greenbaum, S. Hammarling, A. McKenney, and D. Sorensen. *LAPACK Users' Guide, Third Edition.* SIAM, Philadelphia, 1999.
5. R. M. Anderson and R. M. May. *Infectious diseases of humans: dynamics and control.* Oxford University Press, Oxford, UK, 1991.
6. U. Ascher, J. Christiansen, and R. D. Russell. A collocation solver for mixed order systems of boundary value problems. *Math. Computation,* 33:659–679, 1979.
7. U. Ascher, J. Christiansen, and R. D. Russell. Algorithm 569: Colsys: Collocation software for boundary-value ODEs. *ACM Transactions on Mathematical Software (TOMS),* 7(2):223–229, 1981.
8. U. Ascher, J. Christiansen, and R. D. Russell. Collocation software for boundary-value ODEs. *ACM Transactions on Mathematical Software (TOMS),* 7(2):209–222, 1981.
9. J. Baumgarte. Stabilization of constraints and integrals of motion in dynamical systems. *Comp. Math. Appl. Mech. Eng.,* 1972.
10. J. T. Betts. *Practical Methods for Optimal Control Using Nonlinear Programming.* SIAM, 2001.
11. V. Boovaragavan and V. R. Subramanian. A quick and efficient method for consistent initialization of battery models. *Electrochemistry Communications,* 9:1772–1777, 2007.
12. K. E. Brenan, S. L. Campbell, and L. R. Petzold. *The Numerical Solution of Initial Value Problems in Differential-Algebraic Equations.* SIAM, 1996.
13. D. Brown, A. Herbison, J. Robinson, R. Marrs, and G. Leng. Modelling the luteinizing hormone-releasing hormone pulse generator. *Neuroscience,* 63(3):869–879, 1994.
14. P. N. Brown and A. C. Hindmarsh. Reduced storage matrix methods in stiff ODE systems. *J. Appl. Math. Computation,* 31:40–91, 1989.
15. P. N. Brown, A. C. Hindmarsh, and L. R. Petzold. Consistent initial condition calculation for differential-algebraic systems. *SIAM Journal on Scientific Computing,* 1998.
16. C. Bunks, J. Chancelier, F. Delebecque, C. Gomez, M. Goursat, R. Nikoukhah, and S. Steer. *Engineering and Scientific Computing with Scilab.* Birkhäuser, 1999.
17. T. Cambois. Eclipse project: Modeling and simulation of vehicle dynamics and control systems in scicos. In *Proc. of the Scilab Conf.,* Rocquencourt, France, December 2004.
18. S. L. Campbell, J.-P. Chancelier, and R. Nikoukhah. *Modeling and Simulation in Scilab/Scicos.* Springer, New York, New York, 2006.
19. M. Caracotsios and W. E. Stewart. Sensitivity analysis of initial value problems with mixed ODEs and algebraic equations. *Comp. Chem. Eng.,* 9(4):359–365, 1985.

20. J. Chancelier, F. Delebecque, C. Gomez, M. Goursat, R. Nikoukhah, and S. Steer. *Introduction à Scilab*. Springer, 2001.

21. S. D. Cohen and A. C. Hindmarsh. Cvode, a stiff/nonstiff ode solver in c. *Computers in Physics*, 1996.

22. W. R. Cowell, editor. *Sources and Development of Mathematical Software*. Prentice-Hall Series in Computational Mathematics, Cleve Moler, Advisor. Prentice-Hall, Upper Saddle River, NJ 07458, USA, 1984.

23. C. Deboor and R. Weiss. Solveblok: a package for solving almost block diagonal linear systems. *ACM Transactions Mathematical Software*, 6:80–87, 1980.

24. F. Delebecque and R. Nikoukhah. *A mixed symbolic-numeric software environment and its application to control system engineering*, pages 221–244. Elsevier Science Publishers, 1992.

25. E. Demay and J. Bonelle. Metalido : une plate-forme logicielle pour étudier l'exploitation des aménagements hydrauliques. In *Proc. of the Scilab Conf.*, Rocquencourt, France, December 2004.

26. P. Fritzson. *Principles of Object-Oriented Modeling and Simulation with Modelica 2.1*. Wiley-IEEE Press, 2004.

27. M. Goossens, F. Mittelbach, and A. Samarin. *The LaTeX Companion*. Addison Wesley, Reading, Massachusets, 1994.

28. A. Griewank, D. Juedes, H. Mitev, J. Utke, O. Vogel, and A. Walther. ADOL-C: A package for the automatic differentiation of algorithms written in C/C++: Algor. 755. *ACM TOMS*, 22(2):131–167, 1996.

29. S. Guerre-Delabriere and M. Postel. *Méthodes d'approximation - Équations différentielles - Applications Scilab. Niveau L3*. Ellipses, Paris, 2004.

30. E. Hairer, S. P. Norsett, and G. Wanner. *Solving Ordinary Differential Equations I. Nonstiff Problems*, volume 8 of *Springer Series in Comput. Mathematics*. Springer-Verlag, 1993.

31. E. Hairer and G. Wanner. *Solving Ordinary Differential Equations II. Stiff and Differential-Algebraic Problems*, volume 14 of *Springer Series in Comput. Mathematics*. Springer-Verlag, 1991.

32. A. C. Hindmarsh. *ODEPACK, A Systematized Collection of ODE Solvers, in Scientific Computing*. vol. 1 of IMACS Transactions on Scientific Computation. North-Holland, Amsterdam, 1983.

33. A. C. Hindmarsh, P. N. Brown, K. E. Grant, S. L. Lee, R. Serban, D. E. Shumaker, and C. S. Woodward. Sundials: Suite of nonlinear and differential/algebraic equation solvers. *ACM Transactions on Mathematical Software*, 2005.

34. C. V. Hollot and Y. Chait. Nonlinear stability analysis for a class of tcp/aqm networks. In *Proc. of the 40th IEEE Conference on Decision and Control*, pages 2309–2314, Orlando, Florida, December 2001.

35. P. Kunkel and V. Mehrmann. *Differential-Algebraic Equations: Analysis and Numerical Solution*. European Mathematical Society, Zurich, 2006.

36. F. L. Lewis. *Optimal Control*. Wiley, 1986.

37. T. Maly and L. Petzold. Numerical methods and software for sensitivity analysis of differential-algebraic systems. *Applied Numerical Mathematics*, 20:57–79, 1996.

38. L. Mangina, A. Montib, and C. Médigue. Cardiorespiratory system dynamics in chronic heart failure. *European Journal of Heart Failure*, 4:617–625, 2002.

39. S. Mannori and R. Nikoukhah. Scicos: use of hardware in the loop simulation in control. In *Proc. of MESM*, Alexandria, Egypt, August 2006.

40. S. Mannori, R. Nikoukhah, and S. Steer. Commande des procédés industriels avec des logiciels libres. In *Proc. of Journées Démonstrateurs en Automatique*, Angers, France, March 2006.

41. C. Médigue, A. Monti, and A. Wambergue. Lary-cr: Software package for the analysis of cardio vascular and respiratory rhythms, in the scilab-scicos environment. Technical Report 0259, INRIA, April 2002.

42. C. Meza, J. A. Andrade-Romero, R. Bucher, and S. Balemi. Free open source software in control engineering education: A case study in the analysis and control design of a rotary inverted pendulum. In *Proc. of the 14th IEEE International Conference on Emerging Techonologies and Factory Automation*, Mallorca, Spain, September 2009.

43. W. Michiels and S.-I. Niculescu. Stability analysis of a fluid flow model for tcp like behavior. *International Journal of Bifurcation and Chaos*, page to appear, July 2005.

44. M. Najafi and Z. Benjelloun-Dabaghi. Modeling complex systems with modelica in scicos: Application to mean value spark engine. In *Proc. of ESM*, St.Julian's, Malta, October 2007.

45. M. Najafi and Z. Benjelloun-Dabaghi. Modeling and simulation of a drilling station in modelica. In *Proc. of 16th Mediterranean Conference on Control and Automation*, Ajaccio, France, June 2008.

46. M. Najafi, S. Furic, and R. Nikoukhah. Scicos: a general purpose modeling and simulation environment. In *Proc. of 4th International Modelica Conference*, Germany, March 2005.

47. M. Najafi and R. Nikoukhah. Modelica components and toolboxes in scicos. In *Proc. of Scilab Workshop*, Foshan China, May 2007.

48. M. Najafi, X. Shen, Z. Benjelloun-Dabaghi, and P. Moulin. Simulation of the mean-value internal combustion engine in modelica. In *Proc. of 6th MATHMOD*, Vienna, Austria, February 2009.

49. R. Nikoukhah. Extensions to modelica for efficient code generation and separate compilation. In *Proc. of ECOOP*, Berlin, Germany, July 2007.

50. R. Nikoukhah. Hybrid dynamics in modelica: Should all events be considered synchronous. *SNE journal, Special Issues on OO/Hybrid/Structural Dynamic Modeling*, 17, 2008.

51. R. Nikoukhah and S. Furic. Synchronous and asynchronous events in modelica: Proposal for an improved hybrid model. In *Proc. of Modelica Conference*, Bielefeld, Germany, 2008.

52. P. S. M. Pires and D. A. Rogers. Free/open source software: An alternative for engineering students. In *Proc. 32nd ASEE/IEEE Frontiers in Education Conference*, pages T3G-7 TO 11, Boston, MA, 2002.

53. K. Radhakrishnan and A. C. Hindmarsh. Description and use of LSODE, the livermore solver for ordinary differential equations. Technical Report report UCRL-ID-113855, Lawrence Livermore National Laboratory, December 1993.

54. O. E. Rössler. An equation for continuous chaos. *Phys. Lett.*, 35A:397–398, 1976.

55. D. E. Schwarz. A step-by-step approach to compute a consistent initialization for the mna. *International Journal of Circuit Theory and Applications*, 30:1–16, 2002.

Index

Printed in the United States
By Bookmasters